Methods of Molecular Quantum Mechanics

T0201124

Methods of Molecular Quantum Mechanics

An Introduction to Electronic Molecular Structure

Valerio Magnasco
University of Genoa, Genoa, Italy

A John Wiley and Sons, Ltd., Publication

Library of Congress Cataloging-in-Publication Data

Magnasco, Valerio.
   Methods of molecular quantum mechanics : an introduction to electronic molecular structure /
Valerio Magnasco.
       p. cm.
   Includes bibliographical references and index.
   ISBN 978-0-470-68442-9 (cloth) – ISBN 978-0-470-68441-2 (pbk. : alk. paper) 1. Quantum chemistry.
2. Molecular structure. 3. Electrons. I. Title.
   QD462.M335 2009
   541'.28–dc22

                                                                                    2009031405

A catalogue record for this book is available from the British Library.
ISBN H/bk 978-0470-684429 P/bk 978-0470-684412
Set in 10.5/13pt, Sabon by Thomson Digital, Noida, India.
Printed and bound in Great Britain by TJ International Ltd, Padstow, Cornwall.

To my Quantum Chemistry students

Contents

Preface

The structure of this little textbook is essentially methodological and introduces in a concise way the student to a working practice in the *ab initio* calculations of electronic molecular structure, giving a sound basis for a critical analysis of the current calculation programmes. It originates from the need to provide quantum chemistry students with their own personal instant book, giving at low cost a readable introduction to the methods of molecular quantum mechanics, a prerequisite for any understanding of quantum chemical calculations. This book is a recommended companion of the previous book by the author, *Elementary Methods of Molecular Quantum Mechanics*, published in 2007 by Elsevier, which contains many worked examples, and designed as a bridge between Coulson's *Valence* and McWeeny's *Methods of Molecular Quantum Mechanics*. The present book is suitable for a first-year postgraduate university course of about 40 hours.

The book consists of 12 chapters. Particular emphasis is devoted to the Rayleigh variational method, the essential tool for any practical application both in molecular orbital and valence bond theory, and to the stationary Rayleigh–Schroedinger perturbation methods, much attention being given to the Hylleraas variational approximations, which are essential for studying second-order electric properties of molecules and molecular interactions, as well as magnetic properties. In the last chapter, elements on molecular symmetry and group theoretical techniques are briefly presented. Major features of the book are: (i) the consistent use from the very beginning of the system of atomic units (au), essential for simplifying all mathematical formulae; (ii) the introductory use of density matrix techniques for interpreting the properties of many-body systems so as to simplify calculations involving many-electron wavefunctions; (iii) an introduction to valence bond methods, with an explanation of the origin of the chemical bond; and (iv) a unified presentation of basic elements of atomic and molecular interactions, with particular emphasis on the practical use of second-order calculation techniques. Though many examples are treated in depth in this book, for other problems and their detailed solutions the reader may refer to the previous book by the author. The book is completed by alphabetically ordered bibliographical references, and by author and subject indices.

Finally, I wish to thank my son Mario for preparing the drawings at the computer, and my friends and colleagues Deryk W. Davies and Michele Battezzati for their careful reading of the manuscript and useful discussions. In saying that, I regret that, during the preparation of this book, DWD died on 27 February 2008.

I acknowledge support by the Italian Ministry for Education University and Research (MIUR), under grant number 2006 03 0944 003, and Aracne Editrice (Rome) for the 2008 publishing of what is essentially the Italian version entitled *Elementi di Meccanica Quantistica Molecolare*.

Valerio Magnasco
Genoa, 15 May 2009

1

Principles

1.1 THE ORBITAL MODEL

The great majority of the applications of molecular quantum mechanics to chemistry are based on what is called the orbital model. The planetary model of the atom can be traced back to Rutherford (Born, 1962). It consists of a *point-like* nucleus carrying the whole mass and the whole positive charge $+Ze$ surrounded by N electrons each having the elementary negative charge $-e$ and a mass about 2000 times smaller than that of the proton and moving in a space which is essentially that of the atom.[1] Electrons are point-like elementary particles whose negative charge is distributed in space in the form of a *charge cloud*, with the probability of finding the electron at point $\mathbf{r}$ in space being given by

$$|\psi(\mathbf{r})|^2 \, d\mathbf{r} = \text{probability of finding in } d\mathbf{r} \text{ the electron in state } \psi(\mathbf{r})$$

$$(1.1)$$

The functions $\psi(\mathbf{r})$ are called *atomic* orbitals (AOs, one centre) or *molecular* orbitals (MOs, many centres) and describe the quantum states of the electron. For (1.1) to be true, $\psi(\mathbf{r})$ must be a *regular* (or Q-class) mathematical function (single valued, continuous with its first derivatives,

---

[1] The atomic volume has a diameter of the order of 10^2 pm, about 10^5 times larger than that of the nucleus.

Methods of Molecular Quantum Mechanics: An Introduction to Electronic Molecular Structure
Valerio Magnasco
© 2009 John Wiley & Sons, Ltd

quadratically integrable) satisfying the normalization condition

$$\int d\mathbf{r}\, |\psi(\mathbf{r})|^2 = \int d\mathbf{r}\, \psi^*(\mathbf{r})\psi(\mathbf{r}) = 1 \tag{1.2}$$

where integration is extended over the whole space of definition of the variable $\mathbf{r}$ and where $\psi^*(\mathbf{r})$ is the complex conjugate to $\psi(\mathbf{r})$. The last of the above physical constraints implies that ψ must vanish at infinity.[2]

It seems appropriate at this point first to introduce in an elementary way the essential mathematical methods which are needed in the applications, followed by a simple axiomatic formulation of the basic postulates of quantum mechanics and, finally, by their physical interpretation (Margenau, 1961).

1.2   MATHEMATICAL METHODS

In what follows we shall be concerned only with regular functions of the general variable x.

1.2.1   Dirac Notation

$$\begin{cases} \text{Function} & \psi(x) = |\psi\rangle \Rightarrow \text{ket} \\ \text{Complex conjugate} & \psi^*(x) = \langle\psi| \Rightarrow \text{bra} \end{cases} \tag{1.3}$$

The scalar product (see the analogy between regular functions and complex vectors of infinite dimensions) of ψ^* by ψ can then be written in the bra-ket ('bracket') form:

$$\int dx\, \psi^*(x)\psi(x) = \langle\psi|\psi\rangle = \text{finite number} > 0 \tag{1.4}$$

1.2.2   Normalization

If

$$\langle\psi|\psi\rangle = A \tag{1.5}$$

then we say that the function $\psi(x)$ (the ket $|\psi\rangle$) is normalized to A (the *norm* of ψ). The function ψ can then be normalized to 1 by multiplying it by the *normalization factor* $N = A^{-1/2}$.

---

[2] In an atom or molecule, there must be zero probability of finding an electron infinitely far from its nucleus.

1.2.3 Orthogonality

If

$$\langle\psi|\varphi\rangle = \int dx\,\psi^*(x)\varphi(x) = 0 \tag{1.6}$$

then we say that φ is orthogonal ($\perp$) to ψ. If

$$\langle\psi'|\varphi'\rangle = S(\neq 0) \tag{1.7}$$

then φ' and ψ' are *not* orthogonal, but can be orthogonalized by choosing the linear combination (Schmidt orthogonalization):

$$\psi = \psi', \qquad \varphi = N(\varphi' - S\psi'), \qquad \langle\psi|\varphi\rangle = 0 \tag{1.8}$$

where $N = (1 - S^2)^{-1/2}$ is the normalization factor. In fact, it is easily seen that, if ψ' and φ' are normalized to 1:

$$\langle\psi|\varphi\rangle = N\langle\psi'|\varphi' - S\psi'\rangle = N(S - S) = 0 \tag{1.9}$$

1.2.4 Set of Orthonormal Functions

Let

$$\{\varphi_k(x)\} = (\varphi_1\varphi_2\ldots\varphi_k\ldots\varphi_i\ldots) \tag{1.10}$$

be a set of functions. If

$$\langle\varphi_k|\varphi_i\rangle = \delta_{ki} \quad k,i = 1,2,\ldots \tag{1.11}$$

where δ_{ki} is the Kronecker delta (1 if $i = k$, 0 if $i \neq k$), then the set is said to be orthonormal.

1.2.5 Linear Independence

A set of functions is said to be linearly independent if

$$\sum_k \varphi_k(x)C_k = 0 \quad \text{with, necessarily, } C_k = 0 \text{ for any } k \tag{1.12}$$

For a set to be linearly independent, it will be sufficient that the determinant of the metric matrix $\mathbf{M}$ (see Chapter 2) be different from zero:

$$\det \mathbf{M}_{ki} \neq 0 \qquad \mathbf{M}_{ki} = \langle \varphi_k | \varphi_i \rangle \qquad (1.13)$$

A set of orthonormal functions, therefore, is a linearly independent set.

1.2.6 Basis Set

A set of linearly independent functions forms a *basis* in the function space, and we can expand any function of that space into a linear combination of the basis functions. The expansion is unique.

1.2.7 Linear Operators

An operator is a rule transforming a given function into another function (e.g. its derivative). A linear operator $\hat{A}$ satisfies

$$\begin{cases} \hat{A}[\psi_1(x) + \psi_2(x)] = \hat{A}\psi_1(x) + \hat{A}\psi_2(x) \\ \hat{A}[c\psi(x)] = c\hat{A}[\psi(x)] \end{cases} \qquad (1.14)$$

where c is a complex constant. The first and second derivatives are simple examples of linear operators.

1.2.8 Sum and Product of Operators

$$(\hat{A} + \hat{B})\psi(x) = \hat{A}\psi(x) + \hat{B}\psi(x) = (\hat{B} + \hat{A})\psi(x) \qquad (1.15)$$

so that the algebraic sum of two operators is commutative.

In general, the product of two operators is *not* commutative:

$$\hat{A}\hat{B}\psi(x) \neq \hat{B}\hat{A}\psi(x) \qquad (1.16)$$

where the inner operator acts first. If

$$\hat{A}\hat{B} = \hat{B}\hat{A} \qquad (1.17)$$

then the two operators commute. The quantity

$$[\hat{A}, \hat{B}] = \hat{A}\hat{B} - \hat{B}\hat{A} \qquad (1.18)$$

is called the commutator of the operators $\hat{A}, \hat{B}$.

1.2.9   Eigenvalue Equation

The equation

$$\hat{A}\psi(x) = A\psi(x) \tag{1.19}$$

is called the eigenvalue equation for the linear operator $\hat{A}$. When (1.19) is satisfied, the constant A is called the *eigenvalue*, the function ψ the *eigenfunction* of the operator $\hat{A}$. Often, $\hat{A}$ is a differential operator, and there may be a whole spectrum of eigenvalues, each one with its corresponding eigenfunction. The spectrum of the eigenvalues can be either discrete or continuous. An eigenvalue is said to be n-fold degenerate when n different independent eigenfunctions belong to it. We shall see later that the Schroedinger equation for the amplitude $\psi(x)$ is a typical eigenvalue equation, where $\hat{A} = \hat{H} = \hat{T} + V$ is the total energy operator (the Hamiltonian), $\hat{T}$ being the kinetic energy operator and V the potential energy characterizing the system (a scalar quantity).

1.2.10   Hermitian Operators

A Hermitian operator is a linear operator satisfying the so-called 'turn-over rule':

$$\begin{cases} \langle \psi | \hat{A}\varphi \rangle = \langle \hat{A}\psi | \varphi \rangle \\ \int dx\, \psi^*(x)(\hat{A}\varphi(x)) = \int dx\, (\hat{A}\psi(x))^* \varphi(x) \end{cases} \tag{1.20}$$

The Hermitian operators have the following properties:

(i) *real* eigenvalues;
(ii) *orthogonal* (or anyway orthogonalizable) eigenfunctions;
(iii) their eigenfunctions form a *complete* set.

Completeness also includes the eigenfunctions belonging to the continuous part of the eigenvalue spectrum.

Hermitian operators are $-i\partial/\partial x$, $-i\nabla$, $\partial^2/\partial x^2$, ∇^2, $\hat{T} = -(\hbar^2/2m)\nabla^2$ and $\hat{H} = \hat{T} + V$, where i is the imaginary unit ($i^2 = -1$), $\nabla = i(\partial/\partial x) + j(\partial/\partial y) + k(\partial/\partial z)$ is the gradient vector operator, $\nabla^2 = \nabla \cdot \nabla = \partial^2/\partial x^2 + \partial^2/\partial y^2 + \partial^2/\partial z^2$ is the Laplacian operator (in Cartesian coordinates), $\hat{T}$ is the kinetic energy operator for a particle of mass m with $\hbar = h/2\pi$ the reduced Planck constant and $\hat{H}$ is the Hamiltonian operator.

1.2.11   Anti-Hermitian Operators

$\partial/\partial x$ and ∇ are instead anti-Hermitian operators, for which

$$\begin{cases} \left\langle \psi \Big| \dfrac{\partial \varphi}{\partial x} \right\rangle = -\left\langle \dfrac{\partial \psi}{\partial x} \Big| \varphi \right\rangle \\[4mm] \langle \psi | \nabla \varphi \rangle = -\langle \nabla \psi | \varphi \rangle \end{cases} \qquad (1.21)$$

1.2.12   Expansion Theorem

Any regular (Q-class) function $F(x)$ can be expressed exactly in the *complete* set of the eigenfunctions of any Hermitian operator[3]$\hat{A}$. If

$$\hat{A}\varphi_k(x) = A_k \varphi_k(x), \qquad \hat{A}^\dagger = \hat{A} \qquad (1.22)$$

then

$$F(x) = \sum_k \varphi_k(x) C_k \qquad (1.23)$$

where the expansion coefficients are given by

$$C_k = \int dx' \varphi_k^*(x') F(x') = \langle \varphi_k | F \rangle \qquad (1.24)$$

as can be easily shown by multiplying both sides of Equation (1.23) by $\varphi_k^*(x)$ and integrating.

Some authors insert an integral sign into (1.23) to emphasize that integration over the continuous part of the eigenvalue spectrum must be included in the expansion. When the set of functions $\{\varphi_k(x)\}$ is *not* complete, *truncation errors* occur, and a lot of the literature data from the quantum chemistry side is plagued by such errors.

1.2.13   From Operators to Matrices

Using the expansion theorem we can pass from operators (acting on functions) to matrices (acting on vectors; Chapter 2). Consider a *finite*

---

[3] A less stringent stipulation of completeness involves the *approximation in the mean* (Margenau, 1961).

n-dimensional set of basis functions $\{\varphi_k(x)\}k = 1, \ldots, n$. Then, if $\hat{A}$ is a Hermitian operator:

$$\hat{A}\varphi_i(x) = \sum_k \varphi_k(x)A_{ki} = \sum_k |\varphi_k\rangle\langle\varphi_k|\hat{A}\varphi_i\rangle \qquad (1.25)$$

where the expansion coefficients now have *two* indices and are the elements of the square matrix $\mathbf{A}$ (order n):

$$A_{ki} = \langle\varphi_k|\hat{A}\varphi_i\rangle = \int dx'\, \varphi_k^*(x')(\hat{A}\varphi_i(x')) \qquad (1.26)$$

$$\{A_{ki}\} \Rightarrow \mathbf{A} = \begin{pmatrix} A_{11} & A_{12} & \cdots & A_{1n} \\ A_{21} & A_{22} & \cdots & A_{2n} \\ \cdots & \cdots & \cdots & \cdots \\ A_{n1} & A_{n2} & \cdots & A_{nn} \end{pmatrix} = \boldsymbol{\varphi}^\dagger\hat{A}\boldsymbol{\varphi} \qquad (1.27)$$

which is called the *matrix representative* of the operator $\hat{A}$ in the basis $\{\varphi_k\}$, and we use matrix multiplication rules (Chapter 2). In this way, the eigenvalue equations of quantum mechanics transform into eigenvalue equations for the corresponding representative matrices. We must recall, however, that a complete set implies matrices of infinite order.

Under a unitary transformation $\mathbf{U}$ of the basis functions $\boldsymbol{\varphi} = (\varphi_1\varphi_2\cdots\varphi_n)$:

$$\boldsymbol{\varphi}' = \boldsymbol{\varphi}\,\mathbf{U} \qquad (1.28)$$

the representative $\mathbf{A}$ of the operator $\hat{A}$ is changed into

$$\mathbf{A}' = \boldsymbol{\varphi}'^\dagger\hat{A}\boldsymbol{\varphi}' = \mathbf{U}^\dagger\mathbf{A}\mathbf{U} \qquad (1.29)$$

1.2.14 Properties of the Operator ∇

We have seen that in Cartesian coordinates the vector operator ∇ (the gradient, a vector whose components are operators) is defined as (Rutherford, 1962)

$$\nabla = \mathbf{i}\frac{\partial}{\partial x} + \mathbf{j}\frac{\partial}{\partial y} + \mathbf{k}\frac{\partial}{\partial z} \qquad (1.30)$$

Now, let $F(x,y,z)$ be a scalar function of the space point $P(\mathbf{r})$. Then:

$$\nabla F = \mathbf{i}\frac{\partial F}{\partial x} + \mathbf{j}\frac{\partial F}{\partial y} + \mathbf{k}\frac{\partial F}{\partial z} \qquad (1.31)$$

is a vector, the gradient of F.

If $\mathbf{F}$ is a vector of components F_x, F_y, F_z, we then have for the scalar product

$$\nabla \cdot \mathbf{F} = \frac{\partial F_x}{\partial x} + \frac{\partial F_y}{\partial y} + \frac{\partial F_z}{\partial z} = \operatorname{div} \mathbf{F} \qquad (1.32)$$

a scalar quantity, the divergence of $\mathbf{F}$. As a particular case:

$$\nabla \cdot \nabla = \nabla^2 = \frac{\partial^2}{\partial x^2} + \frac{\partial^2}{\partial y^2} + \frac{\partial^2}{\partial z^2} \qquad (1.33)$$

is the Laplacian operator.

From the vector product of ∇ by the vector $\mathbf{F}$ we obtain a new vector, the curl or rotation of $\mathbf{F}$ (written curl F or rot F):

$$\nabla \times \mathbf{F} = \begin{vmatrix} \mathbf{i} & \mathbf{j} & \mathbf{k} \\ \dfrac{\partial}{\partial x} & \dfrac{\partial}{\partial y} & \dfrac{\partial}{\partial z} \\ F_x & F_y & F_z \end{vmatrix} = \operatorname{curl} \mathbf{F} = \mathbf{i}\operatorname{curl}_x \mathbf{F} + \mathbf{j}\operatorname{curl}_y \mathbf{F} + \mathbf{k}\operatorname{curl}_z \mathbf{F} \quad (1.34)$$

a vector operator with components:

$$\begin{cases} \operatorname{curl}_x \mathbf{F} = \dfrac{\partial F_z}{\partial y} - \dfrac{\partial F_y}{\partial z} \\[2mm] \operatorname{curl}_y \mathbf{F} = \dfrac{\partial F_x}{\partial z} - \dfrac{\partial F_z}{\partial x} \\[2mm] \operatorname{curl}_z \mathbf{F} = \dfrac{\partial F_y}{\partial x} - \dfrac{\partial F_x}{\partial y} \end{cases} \qquad (1.35)$$

In quantum mechanics, the vector product of the position vector $\mathbf{r}$ by the linear momentum vector operator $-i\hbar\nabla$ (see Section 1.3) gives the angular momentum vector operator $\hat{\mathbf{L}}$:

$$\hat{\mathbf{L}} = -i\hbar\mathbf{r} \times \nabla = -i\hbar \begin{vmatrix} \mathbf{i} & \mathbf{j} & \mathbf{k} \\ x & y & z \\ \dfrac{\partial}{\partial x} & \dfrac{\partial}{\partial y} & \dfrac{\partial}{\partial z} \end{vmatrix} = \mathbf{i}\hat{L}_x + \mathbf{j}\hat{L}_y + \mathbf{k}\hat{L}_z \qquad (1.36)$$

with components

$$\hat{L}_x = -i\hbar\left(y\frac{\partial}{\partial z} - z\frac{\partial}{\partial y}\right), \quad \hat{L}_y = -i\hbar\left(z\frac{\partial}{\partial x} - x\frac{\partial}{\partial z}\right),$$
$$\hat{L}_z = -i\hbar\left(x\frac{\partial}{\partial y} - y\frac{\partial}{\partial x}\right) \tag{1.37}$$

In the theory of angular momentum, frequent use is made of the *ladder* (or shift) operators:

$$\hat{L}_+ = \hat{L}_x + i\hat{L}_y \text{ (step-up)}, \quad \hat{L}_- = \hat{L}_x - i\hat{L}_y \text{ (step-down)} \tag{1.38}$$

These are also called raising and lowering operators[4] respectively.

Angular momentum operators have the following commutation relations:

$$\begin{cases} [\hat{L}_x, \hat{L}_y] = i\hat{L}_z, \quad [\hat{L}_y, \hat{L}_z] = i\hat{L}_x, \quad [\hat{L}_z, \hat{L}_x] = i\hat{L}_y \\ [\hat{L}_z, \hat{L}_+] = \hat{L}_+, [\hat{L}_z, \hat{L}_-] = -\hat{L}_- \\ [\hat{L}^2, \hat{L}_\kappa] = [\hat{L}^2, \hat{L}_\pm] = 0 \quad \kappa = x, y, z \end{cases} \tag{1.39}$$

The same commutation relations hold for the spin vector operator $\hat{S}$ (Chapter 5).

1.2.15 Transformations in Coordinate Space

We now give the definitions of the main coordinate systems useful in quantum chemistry calculations (Cartesian, spherical, spheroidal), the relations between Cartesian and spherical or spheroidal coordinates, and the expressions of the volume element dr and of the operators ∇ and ∇^2 in the new coordinate systems. We make reference to Figures 1.1 and 1.2.

(i) Cartesian coordinates (x,y,z):

$$x, y, z \in (-\infty, \infty) \tag{1.40}$$

$$\mathbf{dr} = dx\, dy\, dz \tag{1.41}$$

---

[4] Note that the ladder operators are non-Hermitian.

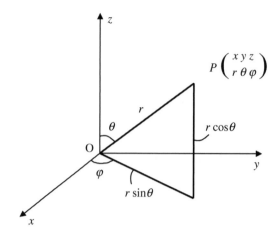

Figure 1.1   Cartesian and spherical coordinate systems

$$\nabla = \mathbf{i}\frac{\partial}{\partial x} + \mathbf{j}\frac{\partial}{\partial y} + \mathbf{k}\frac{\partial}{\partial z} \tag{1.42}$$

$$\nabla^2 = \frac{\partial^2}{\partial x^2} + \frac{\partial^2}{\partial y^2} + \frac{\partial^2}{\partial z^2} \tag{1.43}$$

(ii) Spherical coordinates (r,θ,φ):

$$r(0,\infty),\ \theta(0,\pi),\ \varphi(0,2\pi) \tag{1.44}$$

$$x = r\sin\theta\cos\varphi,\ y = r\sin\theta\sin\varphi,\ z = r\cos\theta \tag{1.45}$$

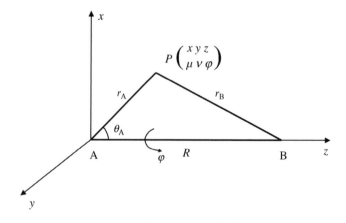

Figure 1.2   Cartesian and spheroidal coordinate systems

$$\mathbf{dr} = r^2 dr \sin\theta \, d\theta \, d\varphi \tag{1.46}$$

$$\nabla = \mathbf{e}_r \frac{\partial}{\partial r} + \mathbf{e}_\theta \frac{1}{r}\frac{\partial}{\partial\theta} + \mathbf{e}_\varphi \frac{1}{r\sin\theta}\frac{\partial}{\partial\varphi} \tag{1.47}$$

$$\nabla^2 = \frac{1}{r^2}\frac{\partial}{\partial r}\left(r^2\frac{\partial}{\partial r}\right) + \frac{1}{r^2}\left[\frac{1}{\sin\theta}\frac{\partial}{\partial\theta}\left(\sin\theta\frac{\partial}{\partial\theta}\right) + \frac{1}{\sin^2\theta}\frac{\partial^2}{\partial\varphi^2}\right]$$

$$= \nabla_r^2 - \frac{\hat{L}^2/\hbar^2}{r^2} \tag{1.48}$$

where $\mathbf{e}_r$, $\mathbf{e}_\theta$, and $\mathbf{e}_\varphi$ are unit vectors along r, θ, and φ. In Equation (1.48):

$$\nabla_r^2 = \frac{1}{r^2}\frac{\partial}{\partial r}\left(r^2\frac{\partial}{\partial r}\right) = \frac{\partial^2}{\partial r^2} + \frac{2}{r}\frac{\partial}{\partial r} \tag{1.49}$$

is the radial Laplacian and

$$\begin{cases} \hat{L}^2 = \hat{\mathbf{L}}\cdot\hat{\mathbf{L}} = -\hbar^2\left[\frac{1}{\sin\theta}\frac{\partial}{\partial\theta}\left(\sin\theta\frac{\partial}{\partial\theta}\right) + \frac{1}{\sin^2\theta}\frac{\partial^2}{\partial\varphi^2}\right] \\[2ex] = -\hbar^2\left(\frac{\partial^2}{\partial\theta^2} + \cot\theta\frac{\partial}{\partial\theta} + \frac{1}{\sin^2\theta}\frac{\partial^2}{\partial\varphi^2}\right) \end{cases} \tag{1.50}$$

is the square of the angular momentum operator (1.36). For the components of the angular momentum vector operator $\hat{L}$ we have

$$\hat{L}_x = -i\hbar\left(-\sin\varphi\frac{\partial}{\partial\theta} - \cot\theta\cos\varphi\frac{\partial}{\partial\varphi}\right) \tag{1.51}$$

$$\hat{L}_y = -i\hbar\left(\cos\varphi\frac{\partial}{\partial\theta} - \cot\theta\sin\varphi\frac{\partial}{\partial\varphi}\right) \tag{1.52}$$

$$\hat{L}_z = -i\hbar\frac{\partial}{\partial\varphi} \tag{1.53}$$

$$\hat{L}_+ = \hbar\exp(i\varphi)\left(\frac{\partial}{\partial\theta} + i\cot\theta\frac{\partial}{\partial\varphi}\right) \tag{1.54}$$

$$\hat{L}_- = \hbar\exp(-i\varphi)\left(-\frac{\partial}{\partial\theta} + i\cot\theta\frac{\partial}{\partial\varphi}\right) \tag{1.55}$$

(iii) Spheroidal coordinates (μ, ν, φ):

$$\mu = \frac{r_A + r_B}{R}(1 \le \mu \le \infty), \quad \nu = \frac{r_A - r_B}{R}(-1 \le \nu \le 1), \quad \varphi(0, 2\pi) \tag{1.56}$$

$$x = \frac{R}{2}\sqrt{(\mu^2 - 1)(1 - \nu^2)}\cos\varphi, \quad y = \frac{R}{2}\sqrt{(\mu^2 - 1)(1 - \nu^2)}\sin\varphi,$$

$$z = \frac{R}{2}(\mu\nu + 1) \tag{1.57}$$

$$d\mathbf{r} = \left(\frac{R}{2}\right)^3 (\mu^2 - \nu^2)\, d\mu\, d\nu\, d\varphi \tag{1.58}$$

$$\nabla = \frac{2}{R}\left[\mathbf{e}_\mu \sqrt{\frac{\mu^2 - 1}{\mu^2 - \nu^2}}\frac{\partial}{\partial\mu} + \mathbf{e}_\nu \sqrt{\frac{1 - \nu^2}{\mu^2 - \nu^2}}\frac{\partial}{\partial\nu} + \mathbf{e}_\varphi \frac{1}{\sqrt{(\mu^2 - 1)(1 - \nu^2)}}\frac{\partial}{\partial\varphi}\right] \tag{1.59}$$

$$\nabla^2 = \frac{4}{R^2(\mu^2 - \nu^2)}$$

$$\times \left\{\frac{\partial}{\partial\mu}\left[(\mu^2 - 1)\frac{\partial}{\partial\mu}\right] + \frac{\partial}{\partial\nu}\left[(1 - \nu^2)\frac{\partial}{\partial\nu}\right] + \frac{\mu^2 - \nu^2}{(\mu^2 - 1)(1 - \nu^2)}\frac{\partial^2}{\partial\varphi^2}\right\} \tag{1.60}$$

Equations (1.44)–(1.55) are used in atomic (one-centre) calculations, whereas Equations (1.56)–(1.60) are used in molecular (at least two-centre) calculations.

1.3 BASIC POSTULATES

We now formulate in an axiomatic way the basis of quantum mechanics in the form of three postulates.

1.3.1 Correspondence between Physical Obervables and Hermitian Operators

In coordinate space, we have the basic correspondences

$$\begin{cases} \mathbf{r} = \mathbf{i}x + \mathbf{j}y + \mathbf{k}z & \Rightarrow \hat{\mathbf{r}} = \mathbf{r} \\ \mathbf{p} = \mathbf{i}p_x + \mathbf{j}p_y + \mathbf{k}p_z & \Rightarrow \hat{\mathbf{p}} = -i\hbar\nabla \end{cases} \tag{1.61}$$

where i is the imaginary unit ($i^2 = -1$) and $\hbar = h/2\pi$ is the reduced Planck constant. More complex observables can be treated by repeated applications of the correspondences (1.61) under the constraint that the resulting quantum mechanical operators must be Hermitian.[5] Kinetic energy and Hamiltonian (*total* energy operator) for a particle of mass m in the potential V are examples already seen. We now give a few further examples by specifying the nature of the potential energy V.

(a) **The one-dimensional harmonic oscillator**
If m is the mass of the oscillator of force constant k, then the Hamiltonian is

$$\hat{H} = -\frac{\hbar^2}{2m}\nabla^2 + \frac{kx^2}{2} \qquad (1.62)$$

(b) **The atomic one-electron problem (the hydrogen-like system)**
If r is the distance of the electron of mass m and charge $-e$ from a nucleus of charge $+Ze$ ($Z = 1$ will give the hydrogen atom), then the Hamiltonian in SI units[6] is

$$\hat{H} = -\frac{\hbar^2}{2m}\nabla^2 - \frac{1}{4\pi\varepsilon_0}\frac{Ze^2}{r} \qquad (1.63)$$

To get rid of all fundamental physical constants in our formulae we shall introduce consistently at this point a system of *atomic units*[7] (au) by posing

$$e = \hbar = m = 4\pi\varepsilon_0 = 1 \qquad (1.64)$$

The basic atomic units of charge, length, energy, and time are expressed in SI units as follows:

$$\begin{cases} \text{charge, } e & e = 1.602\,176\,462 \times 10^{-19}\,\text{C} \\[2mm] \text{length, Bohr} & a_0 = 4\pi\varepsilon_0\dfrac{\hbar^2}{me^2} = 5.291\,772\,087 \times 10^{-11}\,\text{m} \\[3mm] \text{energy, Hartree} & E_h = \dfrac{1}{4\pi\varepsilon_0}\dfrac{e^2}{a_0} = 4.359\,743\,802 \times 10^{-18}\,\text{J} \\[3mm] \text{time} & \tau = \dfrac{\hbar}{E_h} = 2.418\,884\,331 \times 10^{-17}\,\text{s} \end{cases}$$

$$(1.65)$$

---

[5] The quantities observable in physical experiments must be *real*.

[6] An SI dimensional analysis of the two terms of Equation (1.63) shows that they have the dimension of energy (Mohr and Taylor, 2003): $|\hbar^2\nabla^2/2m| = (\text{kg m}^2\,\text{s}^{-1})^2\,\text{m}^{-2}\,\text{kg}^{-1} = \text{kg m}^2\,\text{s}^{-2} = \text{J}$; $|Ze^2/4\pi\varepsilon_0 r| = \text{C}^2\,(\text{J C}^{-2}\,\text{m})\,\text{m}^{-1} = \text{J}$.

[7] Atomic units were first introduced by Hartree (1928a).

At the end of a calculation in atomic units, as we always shall do, the actual SI values can be obtained by taking into account the SI equivalents (1.65).

The Hamiltonian of the hydrogenic system in atomic units will then take the following simplified form:

$$\hat{H} = -\frac{1}{2}\nabla^2 - \frac{Z}{r} \tag{1.66}$$

From now on, we shall consistently use atomic units everywhere, unless explicitly stated.

(c) **The atomic two-electron system**

Two electrons are attracted by a nucleus of charge $+Z$. The Hamiltonian will be

$$\hat{H} = -\frac{1}{2}\nabla_1^2 - \frac{1}{2}\nabla_2^2 - \frac{Z}{r_1} - \frac{Z}{r_2} + \frac{1}{r_{12}} = \hat{h}_1 + \hat{h}_2 + \frac{1}{r_{12}} \tag{1.67}$$

where

$$\hat{h} = -\frac{1}{2}\nabla^2 - \frac{Z}{r} \tag{1.68}$$

is the one-electron Hamiltonian (which has the *same* functional form for both electrons) and the last term is the Coulomb repulsion between the electrons (a two-electron operator). $Z = 2$ gives the He atom.

(d) **The hydrogen molecule-ion H_2^+**

This is a diatomic one-electron molecular system, where the electron is simultaneously attracted by the two protons at A and B. The Born–Oppenheimer Hamiltonian (see Chapter 9) will be

$$\hat{H} = \hat{h} + \frac{1}{R} = -\frac{1}{2}\nabla^2 - \frac{1}{r_A} - \frac{1}{r_B} + \frac{1}{R} = \hat{h}_A + V \tag{1.69}$$

where $\hat{h}_A$ is the one-electron Hamiltonian (1.68) for atom A (with $Z = 1$) and

$$V = -\frac{1}{r_B} + \frac{1}{R} \tag{1.70}$$

is the interatomic potential between the hydrogen atom A and the proton B.

(e) The hydrogen molecule H_2

This is a diatomic two-electron molecular system. The Born–Oppenheimer Hamiltonian will be

$$
\begin{cases}
\hat{H} = \hat{h}_1 + \hat{h}_2 + \dfrac{1}{r_{12}} + \dfrac{1}{R} \\[2mm]
\quad = \left(-\dfrac{1}{2}\nabla_1^2 - \dfrac{1}{r_{A1}} - \dfrac{1}{r_{B1}} \right) + \left(-\dfrac{1}{2}\nabla_2^2 - \dfrac{1}{r_{A2}} - \dfrac{1}{r_{B2}} \right) + \dfrac{1}{r_{12}} + \dfrac{1}{R} \\[2mm]
\quad = \hat{h}_{A1} + \hat{h}_{B2} + V
\end{cases}
\tag{1.71}
$$

where $\hat{h}_{A1}$ and $\hat{h}_{B2}$ are the one-electron Hamiltonians (1.68) for atoms A and B (with $Z = 1$) and

$$
V = -\frac{1}{r_{B1}} - \frac{1}{r_{A2}} + \frac{1}{r_{12}} + \frac{1}{R}
\tag{1.72}
$$

is the interatomic potential between A and B.

1.3.2 State Function and Average Value of Observables

We assume there is a *state function* (or wavefunction, in general complex) $\Psi(x,t)$ that describes in a probabilistic way the dynamical state of a microscopic system. In coordinate space, Ψ is a regular function of coordinate x and time t such that

$$
\begin{aligned}
&\Psi(x,t)\Psi^*(x,t)\,\mathrm{d}x \\
&= \text{probability at time } t \text{ of finding at } \mathrm{d}x \text{ the system in state } \Psi
\end{aligned}
\tag{1.73}
$$

provided Ψ is normalized to 1:

$$
\int \mathrm{d}x\,\Psi^*(x,t)\Psi(x,t) = 1
\tag{1.74}
$$

where integration covers the whole space.

The *average value* of any physical observable[8] A described by the Hermitian operator $\hat{A}$ is obtained from

$$
\langle A \rangle = \frac{\int \mathrm{d}x\,\Psi^*(x,t)\hat{A}\Psi(x,t)}{\int \mathrm{d}x\,\Psi^*(x,t)\Psi(x,t)} = \int \mathrm{d}x\,\hat{A}\,\frac{\Psi(x,t)\Psi^*(x,t)}{\int \mathrm{d}x\,\Psi^*(x,t)\Psi(x,t)}
\tag{1.75}
$$

---

[8] Its expectation value, that can be observed by experiment.

where integration is extended over all space and $\hat{A}$ acts always on Ψ and not on Ψ^*. The last expression above shows that $\hat{A}$ is *weighted* with the (normalized) probability density $\Psi\Psi^*$.

1.3.3   Time Evolution of the State Function

The state function Ψ is obtained by solving the time-dependent Schroedinger equation:

$$\hat{H}\Psi(x,t) = i\hbar\frac{\partial\Psi(x,t)}{\partial t} \qquad (1.76)$$

a partial differential equation which is second order in the space coordinate x and first order in the time t. This equation involves the Hamiltonian of the system $\hat{H}$, so that the total energy E is seen to play a fundamental role among all physical observables.

If the Hamiltonian $\hat{H}$ does not depend explicitly on t (the case of stationary states), then, following the usual mathematical techniques, the variables in Equation (1.76) can be separated by writing Ψ as the product of a space function $\psi(x)$ and a time function $g(t)$:

$$\Psi(x,t) = \psi(x)g(t) \qquad (1.77)$$

giving upon substitution

$$\begin{cases} \hat{H}\psi(x) = E\psi(x) \\ g(t) = g_0\exp(-i\omega t) \end{cases} \qquad (1.78)$$

where E is the separation constant, g_0 is an integration constant, and $\omega = E/\hbar$. The first part of Equation (1.78) is the eigenvalue equation for the total energy operator (the Hamiltonian) of the system, and $\psi(x)$ is called the amplitude function. This is the Schroedinger equation that we must solve or approximate for the physical description of our systems. The second equation gives the time dependence of the stationary state, while general time dependence is fundamental in spectroscopy. It is immediately evident that, for the stationary state, the probability $\Psi\Psi^*\,\mathrm{d}x$ is independent of time:

$$\Psi(x,t)\Psi^*(x,t)\,\mathrm{d}x = |\Psi(x,t)|^2\,\mathrm{d}x = |\psi(x)|^2|g_0|^2\,\mathrm{d}x \qquad (1.79)$$

1.4 PHYSICAL INTERPRETATION OF THE BASIC PRINCIPLES

The explanation of our, so far, unusual postulates is hidden in the nature of the experimental measurements in atomic physics. The experimentally observed *atomicity* of matter (electrons and protons, carrying the elementary negative and positive charge respectively), energy ($h\nu$, Planck), linear momentum (h/λ, De Broglie), and angular momentum ($\hbar$, Bohr) implies some limits in the measurements done at the microscopic level. The direct consequence of the ineliminable interaction between the experimental apparatus and the object of measurement at the subatomic scale was shown by Heisenberg (1930) in his *gedanken Experimente*, and is embodied in his famous uncertainty principle, which for canonically conjugate quantities[9] can be stated in the form

$$\Delta x \Delta p_x \approx h \quad \Delta E \Delta t \approx h \qquad (1.80)$$

where h is the Planck constant, Δx is the uncertainty in the measurement of the x-coordinate, and Δp_x is the uncertainty resulting in the simultaneous measurement of the conjugate linear momentum. It can be seen that the corresponding quantum mechanical operators do not commute:

$$[x, \hat{p}_x] = x\hat{p}_x - \hat{p}_x x = i\hbar \qquad (1.81)$$

whereas operators that are not conjugate commute; say:

$$[x, \hat{p}_y] = x\hat{p}_y - \hat{p}_y x = 0 \qquad (1.82)$$

This means that we cannot measure with the same arbitrary accuracy two conjugated dynamical variables whose quantum mechanical operators do not commute: the exact determination of the position coordinate of the electron would imply the simultaneous infinite inaccuracy in the determination of the corresponding component of the linear momentum!

As a consequence of Heisenberg's principle, the only possible description of the dynamical state of a microscopic body is a *probabilistic* one, and the problem is now to find the function that describes such a probability. This was achieved by Schroedinger (1926a, 1926b, 1926c, 1926d), who assumed that particle matter (the electron) could be

---

[9] In the sense of analytical mechanics.

described by the progressive wave in complex form:

$$\Psi = A \exp(i\alpha) = A \exp[2\pi i(kx - \nu t)] \qquad (1.83)$$

where A is the amplitude and α the phase of a monochromatic plane wave of wavenumber k and frequency ν, which propagates along x. Taking into account the relations of De Broglie and Planck connecting wave-like and particle-like behaviour:

$$k = \frac{p}{h}, \quad \nu = \frac{E}{h} \qquad (1.84)$$

the phase of a *matter wave* could be written as

$$\alpha = \frac{1}{\hbar}(px - Et) \qquad (1.85)$$

giving the wave equation for the matter particle in the form

$$\Psi = A \exp\left[\frac{i}{\hbar}(px - Et)\right] = \Psi(x, t) \qquad (1.86)$$

which defines Ψ as a function of x and t at constant values of p and E. Then, taking the derivatives of Ψ with respect to x and t, we obtain respectively

$$\frac{\partial \Psi}{\partial x} = \frac{i}{\hbar}p\Psi \quad \text{whence} \Rightarrow \hat{p} = -i\hbar\frac{\partial}{\partial x} \qquad (1.87)$$

which is our first postulate (1.61) for the x-component of the linear momentum, and

$$\frac{\partial \Psi}{\partial t} = -\frac{i}{\hbar}E\Psi \quad \text{whence} \Rightarrow \hat{E} = \hat{H} = i\hbar\frac{\partial}{\partial t} \qquad (1.88)$$

the time-dependent Schroedinger Equation (1.76) giving the time evolution of the state function Ψ. Hence, the two correspondences (1.87) and (1.88), connecting linear momentum and total energy to the first derivatives of the function Ψ, necessarily imply the fundamental relations occurring in our axiomatic proposition of quantum mechanics.

As a consequence of our probabilistic description, in doing experiments in atomic physics we usually obtain a *distribution* of the observable eigenvalues, unless the state function Ψ coincides at time t with the eigenfunction φ_κ of the corresponding quantum mechanical operator, in which case we have a 100% probability of observing for A the value A_κ. Such probability distributions fluctuate in time for all observables but the energy, where we have a distribution of eigenvalues (the possible values of

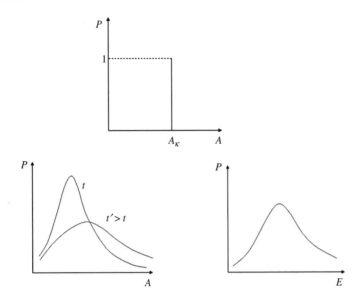

Figure 1.3 Probability distribution of the physical observable A. Top row: definite value for the κth eigenvalue. Bottom row: fluctuation in time of the probability distribution (left) and distribution of energy eigenvalues constant in time (right)

the energy levels) constant in time. These different cases are qualitatively depicted in the plots of Figure 1.3.

The formulation of quantum mechanics of Section 1.3 is the most useful for us, but not the only one possible. For instance, using Dirac's approach, Troup (1968) showed the equivalence between our Schroedinger[10] 'wave mechanics' (the *continuous* approach) and Heisenberg 'matrix mechanics' (the *discrete* approach), giving working examples in both cases (e.g. the harmonic oscillator and the hydrogen atom). However, this matter is outside the scope of the present book.

---

[10] Based on the Schroedinger (1926a, 1926b, 1926c, 1926d) series of four papers entitled 'Quantisierung als Eigenwertproblem'. The equivalence between matrix and wave mechanics was examined by Schroedinger (1926e) in an intermediate paper.

2

Matrices

Matrices are the powerful algorithm connecting the differential equations of quantum mechanics to equations governed by the linear algebra of matrices and their transformations. After a short introduction on elementary properties of matrices and determinants (Margenau and Murphy, 1956; Aitken, 1958; Hohn, 1964), we introduce special matrices and the matrix eigenvalue problem.

2.1 DEFINITIONS AND ELEMENTARY PROPERTIES

A matrix $\mathbf{A}$ of *order* $m \times n$ is an array of numbers or functions ordered according to m rows and n columns:

$$\mathbf{A} = \begin{pmatrix} A_{11} & A_{12} & \cdots & A_{1n} \\ A_{21} & A_{22} & \cdots & A_{2n} \\ \cdots & \cdots & \cdots & \cdots \\ A_{m1} & A_{m2} & \cdots & A_{mn} \end{pmatrix} \tag{2.1}$$

and can be denoted by its ij element $(i = 1, 2, \ldots, m; j = 1, 2, \ldots, n)$ as

$$\mathbf{A} = \{A_{ij}\} \tag{2.2}$$

Matrix $\mathbf{A}$ is rectangular if $n \neq m$ and square if $n = m$. In square matrices, elements with $j = i$ are called diagonal. To any square matrix $\mathbf{A}$ we can associate two scalar quantities: its *determinant* $|A| = \det A$ (a number) and its *trace* $\operatorname{tr} \mathbf{A} = \sum_i A_{ii}$, the sum of all diagonal elements.

Methods of Molecular Quantum Mechanics: An Introduction to Electronic Molecular Structure
Valerio Magnasco
© 2009 John Wiley & Sons, Ltd

Two matrices **A** and **B** of the same order are equal if

$$\mathbf{B} = \mathbf{A} \quad B_{ij} = A_{ij} \text{ for all } i, j \qquad (2.3)$$

Matrices can be added (or subtracted) if they have the same order:

$$\mathbf{A} \pm \mathbf{B} = \mathbf{C} \quad C_{ij} = A_{ij} \pm B_{ij} \qquad (2.4)$$

Addition and subtraction enjoy commutative and associative properties.

Multiplying a matrix **A** by a complex number c implies multiplication of *all* elements of **A** by that number:

$$c\mathbf{A} = \mathbf{B} \quad B_{ij} = cA_{ij} \qquad (2.5)$$

The product, rows by columns, of two (or more) matrices **A** by **B** is possible if the matrices are conformable (the number of columns of **A** equals the number of rows of **B**):

$$\underset{m\times n \; n\times p \; m\times p}{\mathbf{AB} = \mathbf{C}} \quad C_{ij} = \sum_{\alpha=1}^{n} A_{i\alpha} B_{\alpha j}, \quad \underset{m\times n \; n\times p \; p\times q \; m\times q}{\mathbf{ABC} = \mathbf{D}},$$

$$D_{ij} = \sum_{\alpha=1}^{n} \sum_{\beta=1}^{p} A_{i\alpha} B_{\alpha\beta} C_{\beta j} \qquad (2.6)$$

Matrix multiplication is usually *not* commutative, the quantity

$$[\mathbf{A}, \mathbf{B}] = \mathbf{AB} - \mathbf{BA} \qquad (2.7)$$

being the *commutator* of **A** and **B**. If

$$[\mathbf{A}, \mathbf{B}] = 0 \qquad (2.8)$$

then matrices **A** and **B** commute.

The product of more than two matrices enjoys the associative property:

$$\mathbf{ABC} = (\mathbf{AB})\mathbf{C} = \mathbf{A}(\mathbf{BC}) \qquad (2.9)$$

The trace of a product of matrices is invariant under the cyclic permutation of its factors.

2.2 PROPERTIES OF DETERMINANTS

Given a determinant $|\mathbf{A}|$ of order n (n rows and n columns), we call $|A_{ij}|$ the *minor* of $|\mathbf{A}|$, a determinant of order $(n-1)$ obtained from $\mathbf{A}$ by deleting row i and column j, and by $|a_{ij}| = (-1)^{i+j}|A_{ij}|$ the *cofactor* (signed minor) of $|\mathbf{A}|$.

The main properties of determinants will be briefly recalled here (Aitken, 1958).

1. A determinant can be expanded in an elementary way in terms of any of its rows or columns:

$$|\mathbf{A}| = \begin{vmatrix} A_{11} & A_{12} & \cdots & A_{1n} \\ A_{21} & A_{22} & \cdots & A_{2n} \\ \cdots & \cdots & \cdots & \cdots \\ A_{n1} & A_{n2} & \cdots & A_{nn} \end{vmatrix} = \sum_{j=1}^{n} A_{ij}|a_{ij}| = \sum_{i=1}^{n} A_{ij}|a_{ij}| \quad (2.10)$$

  the first being the expansion according to row i and the second according to column j. The expansion of a determinant of order n gives $n!$ terms.
2. A determinant changes sign if we interchange any two rows or columns.
3. A determinant is unchanged upon interchanging all rows and columns.
4. The product of $|\mathbf{A}|$ by a complex number c implies that only *one* row (or column) is multiplied by that number.
5. A determinant vanishes if two rows or two columns are identical.
6. A determinant vanishes if all elements of a row or column are zero.
7. If each element in any one row (or column) is the sum of two or more quantities, then the determinant can be written as the sum of two or more determinants of the same order (distribution property). For instance: $|a+bc| = |ac| + |bc|$, where for short we write only the diagonal elements.
8. The determinant of the product of two or more square matrices of the same order is the product of the determinants of the individual matrices:

$$|\mathbf{ABC}| = |\mathbf{A}| \cdot |\mathbf{B}| \cdot |\mathbf{C}| \quad (2.11)$$

2.3 SPECIAL MATRICES

1. The null matrix (even rectangular, all elements are zero):

$$\mathbf{0} = \begin{pmatrix} 0 & 0 & \cdots & 0 \\ 0 & 0 & \cdots & 0 \\ \cdots & \cdots & \cdots & \cdots \\ 0 & 0 & \cdots & 0 \end{pmatrix} \tag{2.12}$$

$$\mathbf{0} + \mathbf{A} = \mathbf{A} + \mathbf{0} = \mathbf{A} \quad \mathbf{A0} = \mathbf{0A} = \mathbf{0} \quad (\textit{same } \text{order}) \tag{2.13}$$

2. The diagonal matrix (order n):

$$\mathbf{\Lambda} = \begin{pmatrix} \lambda_1 & 0 & \cdots & 0 \\ 0 & \lambda_2 & \cdots & 0 \\ \cdots & \cdots & \cdots & \cdots \\ 0 & 0 & \cdots & \lambda_n \end{pmatrix} \qquad \lambda_{ij} = \lambda_i \delta_{ij} \tag{2.14}$$

where δ_{ij} is the Kronecker delta.

3. The identity matrix (order n):

$$\mathbf{1} = \begin{pmatrix} 1 & 0 & \cdots & 0 \\ 0 & 1 & \cdots & 0 \\ \cdots & \cdots & \cdots & \cdots \\ 0 & 0 & \cdots & 1 \end{pmatrix} \qquad 1_{ij} = \delta_{ij} \tag{2.15}$$

$$\mathbf{1A} = \mathbf{A1} = \mathbf{A} \quad (\textit{same } \text{order}) \tag{2.16}$$

4. The scalar matrix (order n):

$$\mathbf{\Lambda} = \begin{pmatrix} \lambda & 0 & \cdots & 0 \\ 0 & \lambda & \cdots & 0 \\ \cdots & \cdots & \cdots & \cdots \\ 0 & 0 & \cdots & \lambda \end{pmatrix} = \lambda \mathbf{1} \tag{2.17}$$

5. Given a square matrix $\mathbf{A}$ of order n, we can construct the following matrices:

$$\begin{cases} \mathbf{B} = \mathbf{A}^* & B_{ij} = A_{ij}^* & \text{complex conjugate} \\ \mathbf{B} = \tilde{\mathbf{A}} & B_{ij} = A_{ji} & \text{transpose} \\ \mathbf{B} = \mathbf{A}^\dagger & B_{ij} = A_{ji}^* & \text{adjoint} \\ \mathbf{B} = \mathbf{A}^{-1} & B_{ij} = (\det A)^{-1}|a_{ji}| & \text{inverse} \end{cases} \qquad (2.18)$$

When performed twice, the operations $*$, $\sim$, $\dagger$, and -1 restore the original matrix. Products have the properties

$$(\mathbf{AB})^\sim = \tilde{\mathbf{B}}\tilde{\mathbf{A}} \quad (\mathbf{AB})^\dagger = \mathbf{B}^\dagger\mathbf{A}^\dagger \quad (\mathbf{AB})^{-1} = \mathbf{B}^{-1}\mathbf{A}^{-1} \qquad (2.19)$$

6. Given a square matrix $\mathbf{A}$ of order n, if

$$\mathbf{A} = \mathbf{A}^*, \quad \mathbf{A} = \tilde{\mathbf{A}}, \quad \text{or} \quad \mathbf{A} = \mathbf{A}^\dagger \qquad (2.20)$$

then we say that $\mathbf{A}$ is real, symmetric, or Hermitian (or self-adjoint) respectively.

7. If

$$\mathbf{A}^{-1} = \tilde{\mathbf{A}} \quad \mathbf{A}^{-1} = \mathbf{A}^\dagger \qquad (2.21)$$

then we say that $\mathbf{A}$ is orthogonal or unitary respectively.

2.4 THE MATRIX EIGENVALUE PROBLEM

A system of linear inhomogeneous algebraic equations in the n unknowns c_i $(i = 1, 2, \ldots, n)$ can be written in matrix form as

$$\mathbf{Ac} = \mathbf{b} \qquad (2.22)$$

if we introduce the matrices

$$\mathbf{A} = \begin{pmatrix} A_{11} & A_{12} & \cdots & A_{1n} \\ A_{21} & A_{22} & \cdots & A_{2n} \\ \cdots & \cdots & \cdots & \cdots \\ A_{n1} & A_{n2} & \cdots & A_{nn} \end{pmatrix} \qquad (2.23)$$

(i.e. the square matrix of coefficients)

$$\mathbf{c} = \begin{pmatrix} c_1 \\ c_2 \\ \cdots \\ c_n \end{pmatrix} \qquad (2.24)$$

(i.e. the column vector of the unknowns)

$$\mathbf{b} = \begin{pmatrix} b_1 \\ b_2 \\ \cdots \\ b_n \end{pmatrix} \qquad (2.25)$$

(i.e. the column vector of the inhomogeneous terms), and adopt matrix multiplication rules.

Matrix equation (2.22) can be interpreted as a linear transformation on vector $\mathbf{c}$, which is transformed into vector $\mathbf{b}$ under the action of matrix $\mathbf{A}$. If $\mathbf{A}^{-1}$ exists ($\det A \neq 0$), then the solution of system (2.22) is given by

$$\mathbf{c} = \mathbf{A}^{-1}\mathbf{b} \qquad (2.26)$$

which is nothing but the well-known Cramer's rule.

When $\mathbf{b}$ is proportional to $\mathbf{c}$ through a number λ, then

$$\mathbf{Ac} = \lambda\mathbf{c} \qquad (2.27)$$

and we obtain what is known as the *eigenvalue equation* for the square matrix $\mathbf{A}$. By writing

$$(\mathbf{A} - \lambda 1)\mathbf{c} = 0 \qquad (2.28)$$

we obtain a system of linear homogeneous algebraic equations in the unknowns $\mathbf{c}$, which has nontrivial solutions if and only if the determinant of the coefficients vanishes:

$$\det(\mathbf{A} - \lambda 1) = \begin{vmatrix} A_{11} - \lambda & A_{12} & \cdots & A_{1n} \\ A_{21} & A_{22} - \lambda & \cdots & A_{2n} \\ \cdots & \cdots & \cdots & \cdots \\ A_{n1} & A_{n2} & \cdots & A_{nn} - \lambda \end{vmatrix} = 0 \qquad (2.29)$$

Equation (2.29) is known as characteristic (or *secular*) equation of the square matrix $\mathbf{A}$. It is an algebraic equation of degree n in λ, with

$$\begin{cases} \lambda_1, \lambda_2, \ldots, \lambda_n & n \text{ roots (the } eigenvalues \text{ of } \mathbf{A}) \\ \mathbf{c}_1, \mathbf{c}_2, \ldots, \mathbf{c}_n & n \text{ column coefficients (the } eigenvectors \text{ of } \mathbf{A}) \end{cases} \quad (2.30)$$

The whole set of the n eigenvalue equations for $\mathbf{A}$

$$\mathbf{Ac}_1 = \lambda_1 \mathbf{c}_1, \quad \mathbf{Ac}_2 = \lambda_2 \mathbf{c}_2, \quad \ldots, \quad \mathbf{Ac}_n = \lambda_n \mathbf{c}_n \quad (2.31)$$

can be replaced by the *full* eigenvalue equation

$$\mathbf{AC} = \mathbf{C\Lambda} \quad (2.32)$$

if we introduce the following square matrices of order n:

$$\mathbf{\Lambda} = \begin{pmatrix} \lambda_1 & 0 & \cdots & 0 \\ 0 & \lambda_2 & \cdots & 0 \\ \cdots & \cdots & \cdots & \cdots \\ 0 & 0 & \cdots & \lambda_n \end{pmatrix},$$

$$\mathbf{C} = (\mathbf{c}_1 \mathbf{c}_2 \cdots \mathbf{c}_n) = \begin{pmatrix} c_{11} & c_{12} & \cdots & c_{1n} \\ c_{21} & c_{22} & \cdots & c_{2n} \\ \cdots & \cdots & \cdots & \cdots \\ c_{n1} & c_{n2} & \cdots & c_{nn} \end{pmatrix} \quad (2.33)$$

where $\mathbf{\Lambda}$ is the diagonal matrix of the n eigenvalues and $\mathbf{C}$ is the row matrix of the n eigenvectors, a square matrix on the whole.

If $\det \mathbf{C} \neq 0$, then $\mathbf{C}^{-1}$ exists and the square matrix $\mathbf{A}$ can be brought to diagonal form through the transformation

$$\mathbf{C}^{-1}\mathbf{AC} = \mathbf{\Lambda} \quad (2.34)$$

a process which is called the *diagonalization* of matrix $\mathbf{A}$.

If $\mathbf{A}$ is Hermitian

$$\mathbf{A} = \mathbf{A}^\dagger \quad (2.35)$$

then eigenvalues are *real* and eigenvectors *orthogonal*, so that the complete matrix of the eigenvectors is a unitary matrix $(\mathbf{C}^{-1} = \mathbf{C}^\dagger)$, and (2.34) therefore becomes:

$$\mathbf{C}^{\dagger}\mathbf{AC} = \mathbf{\Lambda} \tag{2.36}$$

Namely, a Hermitian matrix $\mathbf{A}$ can be brought to diagonal form by a unitary transformation with the complete matrix of its eigenvectors.

We examine below the simple case of the 2×2 Hermitian matrix $\mathbf{A}$:

$$\mathbf{A} = \begin{pmatrix} 1 & S \\ S & 1 \end{pmatrix} \tag{2.37}$$

The secular equation is

$$\begin{vmatrix} 1-\lambda & S \\ S & 1-\lambda \end{vmatrix} = 0 \tag{2.38}$$

giving upon expansion the quadratic equation in λ:

$$\lambda^2 - 2\lambda + 1 - S^2 = 0 \tag{2.39}$$

with the roots (the eigenvalues):

$$\lambda_1 = 1 - S, \quad \lambda_2 = 1 + S \tag{2.40}$$

We now turn to the evaluation of the eigenvectors.

(i) $\lambda_1 = 1 - S$

$$\begin{cases} (1-\lambda_1)c_1 + Sc_2 = 0 \\ c_1^2 + c_2^2 = 1 \end{cases} \tag{2.41}$$

We solve the homogeneous linear system (2.28) for the *first* eigenvalue with the additional constraint of coefficients normalization:[1]

$$\left(\frac{c_2}{c_1}\right)_1 = \frac{\lambda_1 - 1}{S} = -\frac{S}{S} = -1 \Rightarrow c_2 = -c_1 \tag{2.42}$$

---

[1] Solution of the homogeneous system is seen to give only the ratio c_2/c_1.

$$c_1^2 + c_2^2 = 2c_1^2 = 1 \Rightarrow c_1 = \frac{1}{\sqrt{2}}, \quad c_2 = -\frac{1}{\sqrt{2}} \qquad (2.43)$$

$$\mathbf{c_1} = \begin{pmatrix} \dfrac{1}{\sqrt{2}} \\ -\dfrac{1}{\sqrt{2}} \end{pmatrix} \qquad (2.44)$$

(ii) $\lambda_2 = 1 + S$

$$\begin{cases} (1 - \lambda_2)c_1 + Sc_2 = 0 \\ c_1^2 + c_2^2 = 1 \end{cases} \qquad (2.45)$$

We now solve system (2.28) for the *second* eigenvalue with the additional constraint of coefficients normalization:

$$\left(\frac{c_2}{c_1}\right)_2 = \frac{\lambda_2 - 1}{S} = \frac{S}{S} = 1 \Rightarrow c_2 = c_1 \qquad (2.46)$$

$$c_1^2 + c_2^2 = 2c_1^2 = 1 \Rightarrow c_1 = c_2 = \frac{1}{\sqrt{2}} \qquad (2.47)$$

$$\mathbf{c_2} = \begin{pmatrix} \dfrac{1}{\sqrt{2}} \\ \dfrac{1}{\sqrt{2}} \end{pmatrix} \qquad (2.48)$$

It is left as an easy exercise for the reader to verify by direct matrix multiplication that the complete matrix $\mathbf{C}$ of the eigenvectors

$$\mathbf{C} = (\mathbf{c_1}\, \mathbf{c_2}) = \begin{pmatrix} \dfrac{1}{\sqrt{2}} & \dfrac{1}{\sqrt{2}} \\ -\dfrac{1}{\sqrt{2}} & \dfrac{1}{\sqrt{2}} \end{pmatrix} \qquad (2.49)$$

is a unitary matrix

$$\mathbf{C}^\dagger \mathbf{C} = \mathbf{C}\mathbf{C}^\dagger = 1 \qquad (2.50)$$

and that

$$C^{\dagger}AC = \Lambda \qquad (2.51)$$

More advanced techniques for the calculation of functions of Hermitian matrices and matrix projectors can be found elsewhere (Magnasco, 2007).

3

Atomic Orbitals

3.1 ATOMIC ORBITALS AS A BASIS FOR MOLECULAR CALCULATIONS

We saw in Chapter 1 that AOs are one-electron one-centre functions needed for describing the probability of finding the electron at any given point in space. They are, therefore, the building blocks of any theory that can be devised inside the orbital model. In practical applications, we shall see in Chapter 4 that appropriate orbitals will be the basis of all approximation methods resting on the variation theorem. A particular type of AO is obtained from the solution of the atomic one-electron problem, the so-called hydrogen-like atomic orbitals (HAOs). Even if the HAOs are of no interest in practice, they are important in that they are *exact* solutions of the corresponding Schroedinger equation and, therefore, are useful for testing the accuracy of approximate calculations. The great majority of quantum chemical calculations on atoms and molecules are based on the use of basis AOs that have a radial dependence different from that of the HAOs. They can be separated into two classes according to whether their decay with the radial variable r is exponential (Slater-type orbitals or STOs, by far the best) or Gaussian (Gaussian-type orbitals or GTOs).

In the following, we shall first introduce the HAOs mostly with the aim of (i) illustrating the general techniques of solution of one of the exactly solvable Schroedinger eigenvalue equations and (ii) explaining from first

Methods of Molecular Quantum Mechanics: An Introduction to Electronic Molecular Structure
Valerio Magnasco
© 2009 John Wiley & Sons, Ltd

principles the origin of the quantum numbers (n, ℓ, and m) that characterize these orbitals, and which arise from the regularity conditions imposed upon the mathematical solutions. We shall move next to consideration of STOs and GTOs, giving some general definitions and some simple one-centre one-electron integrals which will be needed in Chapter 4.

3.2   HYDROGEN-LIKE ATOMIC ORBITALS

HAOs are obtained as *exact* solutions of the Schroedinger eigenvalue equation for the atomic one-electron system:

$$\hat{H}\psi = E\psi \quad \hat{H} = -\frac{1}{2}\nabla^2 - \frac{Z}{r} \tag{3.1}$$

where Z is the nuclear charge, giving for $Z = 1, 2, 3, 4, \ldots$ the isoelectronic series H, He^+, Li^{2+}, Be^{3+}, ... In what follows we shall go briefly through the solution of the eigenvalue equation, Equation 3.1. Solution of the second-order *partial* differential equation embodied in (3.1) necessarily involves the steps outlined in Sections 3.2.1–3.2.5.

3.2.1   Choice of an Appropriate Coordinate System

The spherical symmetry of the potential energy $V(r)$ suggests use of the spherical coordinates (r, θ, φ) (see Figure 1.1). We have seen that in this case the Laplacian operator ∇^2 separates into a *radial* Laplacian ∇_r^2 and into an *angular* part which depends on the square of the angular momentum operator $\hat{L}^2$. In this way, it is possible to separate radial from angular equations if we put

$$\psi(r, \theta, \varphi) = R(r)Y(\theta, \varphi) \tag{3.2}$$

Upon substitution in (3.1) we obtain the following two separate differential equations:

$$\frac{d^2R}{dr^2} + \frac{2}{r}\frac{dR}{dr} + \left[2\left(E + \frac{Z}{r}\right) - \frac{\lambda}{r^2}\right]R = 0 \tag{3.3}$$

$$\hat{L}^2 Y = \lambda Y \tag{3.4}$$

where $\lambda \geq 0$ is a first separation constant.[1] Equation 3.3 is the differential equation determining the *radial* part of the HAOs and Equation 3.4 is the eigenvalue equation for the square of the angular momentum operator $\hat{L}^2$ determining the *angular* part of the orbitals. The latter equation is found in general in the study of potential theory, the eigenfunctions $Y(\theta, \varphi)$ in complex form being known in mathematics as *spherical harmonics*. At variance with the radial eigenfunctions $R(r)$, which are peculiar to the hydrogen-like system, the $Y(\theta, \varphi)$ are useful in general for atoms.

3.2.2 Solution of the Radial Equation

The radial equation (3.3) has different solutions according to the value of the parameter E, the eigenvalue of Equation 3.1.

$E > 0$ corresponds to the *continuous* spectrum of the ionized atom, its eigenfunctions being oscillatory solutions describing plane waves. It is of no interest to us here, except for completing the spectrum of the Hermitian operator $\hat{H}$.

$E < 0$ corresponds to the electron bound to the nucleus, with a *discrete* spectrum of eigenvalues, the energy levels of the hydrogen-like atom as observed from atomic spectra.

It is customary to pose $R(r) = P(r)/r$ and

$$E = -\frac{Z^2}{2n^2} \tag{3.5}$$

where n is a real integer positive parameter to be determined,[2] and to change the variable to

$$x = \frac{Z}{n} r \tag{3.6}$$

---

[1] We shall see later in this section that $\lambda = \ell(\ell + 1)$, where $\ell \geq 0$ is the orbital quantum number.

[2] It is seen that (3.5) is nothing but the result in atomic units of Bohr's theory for the hydrogenic system of nuclear charge $+Z$.

Therefore, the second-order *ordinary* differential equation (3.3) becomes

$$\frac{d^2 P}{dx^2} + \left[-1 + \frac{2n}{x} - \frac{\ell(\ell+1)}{x^2} \right] P = 0 \tag{3.7}$$

which must be solved with its regularity conditions in the interval $0 \leq x \leq \infty$ including the extrema.

The study of the asymptotic behaviour of the function $P(x)$ for the electron *far* from or *near* to the nucleus shows that $P(x)$ must have the form

$$P(x) = \exp(-x) x^{\ell+1} F(x) \tag{3.8}$$

where $F(x)$ is a function to be determined from the solution of the differential equation

$$x \frac{d^2 F}{dx^2} + [(2\ell+2) - 2x] \frac{dF}{dx} + 2(n - \ell - 1)F = 0 \tag{3.9}$$

Equation 3.9 is solved by a power series expansion (Taylor) in the x variable:

$$F(x) = \sum_k a_k x^k \tag{3.10}$$

Upon substitution in (3.9), a two-term recursion formula is obtained for the coefficients a_k:

$$a_{k+1} = \frac{2(k - n + \ell + 1)}{(k+1)(k+2\ell+2)} a_k \quad k = 0, 1, 2, \dots \tag{3.11}$$

The study of the convergence of the power *series* (Equation 3.10) shows that it converges to the function $\exp(2x)$. This solution is not physically acceptable since, in this case, $P(x)$ would diverge at $x = \infty$. So, the regularity conditions on the radial function $P(x)$ require that the expansion Equation 3.10 should be truncated to a *polynomial*. This can be achieved if

$$a_k \neq 0, \quad a_{k+1} = a_{k+2} = \cdots = 0 \tag{3.12}$$

which implies from (3.11) the necessary condition

$$k - n + \ell + 1 = 0 \quad \Rightarrow \quad k_{\max} = n - \ell - 1 \tag{3.13}$$

The physically acceptable radial solutions must, hence, include a polynomial of degree $(n - \ell - 1)$ at most.[3] Equation 3.13 determines our, so far, unknown parameter n:

$$n = k + \ell + 1 \quad \Rightarrow \quad n = \ell + 1, \ell + 2, \ldots \quad \Rightarrow n \geq \ell + 1 \tag{3.14}$$

and, in this way, we obtain the well-known relation between principal quantum number n and orbital quantum number ℓ.

Coming back to $R(x) = P(x)/x$, we see that our radial functions $R(x)$ will depend on quantum numbers n and ℓ, being written in un-normalized form as

$$R_{n\ell}(x) = \exp(-x)x^{\ell} \sum_{k=0}^{n-\ell-1} a_k x^k \tag{3.15}$$

with

$$n = 1, 2, 3, 4, \ldots \quad \ell = 0, 1, 2, 3, \ldots, (n-1) \tag{3.16}$$

The functions in Equation 3.15 have $(n - \ell - 1)$ nodes, namely those values of x for which the function changes sign. The detailed form of the first few radial functions can be readily obtained from (3.15) and (3.16) using the recursion formula (3.11) for the coefficients. It is found that

---

[3] Mathematically speaking, these are related to the associated Laguerre polynomials $L_{n+\ell}^{2\ell+1}(x)$ (Eyring *et al.*, 1944).

$$\begin{cases} n=1, \ell=0 \quad R_{10} = \exp(-x)a_0 \\[8pt] n=2, \ell=0 \quad R_{20} = \exp(-x)(a_0+a_1x) = \exp(-x)(1-x)a_0 \\[8pt] n=3, \ell=0 \quad R_{30} = \exp(-x)(a_0+a_1x+a_2x^2) \qquad \qquad ns\ \text{functions} \\[8pt] \qquad \qquad \quad = \exp(-x)\left(1-2x+\frac{2}{3}x^2\right)a_0 \end{cases}$$

$$(3.17)$$

$$\begin{cases} n=2, \ell=1 \quad R_{21} = \exp(-x)xa_0 \\[8pt] n=3, \ell=1 \quad R_{31} = \exp(-x)x(a_0+a_1x) \\[8pt] \qquad \qquad \quad = \exp(-x)x\left(1-\frac{1}{2}x\right)a_0 \end{cases} \quad np\ \text{functions} \qquad (3.18)$$

$$n=3, \ell=2 \qquad R_{32} = \exp(-x)x^2a_0 \quad \text{3d function} \qquad (3.19)$$

where a_0 is a normalization factor. The plots of $R_{n\ell}(x)$ versus x for 1s, 2s, 2p, 3d HAOs are given in Figure 3.1. It is seen that the functions with $\ell=0$

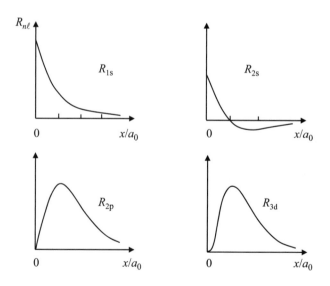

Figure 3.1 $R_{n\ell}$ radial functions for 1s, 2s, 2p, and 3d HAOs

(called s) all have a *cusp* at the origin, while the functions with $\ell = 1$(p), $\ell = 2$(d), $\ell = 3$(f), ... are all *zero* at the origin, but with a different behaviour of their first derivative there. It is recommended that the reader verify that such radial functions are the correct solutions of the differential equation (3.7) with $P(x) = xR(x)$.

3.2.3 Solution of the Angular Equation

We shall now give a short glance to the solution of the angular equation (3.4), which follows much the same lines of solution as we have just seen for the radial equation. The first step is the separation of the partial differential equation in the angular variables θ and φ into two differential equations, one for each variable, by posing

$$Y(\theta, \varphi) = \Theta(\theta)\Phi(\varphi) \tag{3.20}$$

This implies a second separation constant, which will for convenience be called m^2. Of the two resultant total differential equations

$$\frac{d^2\Phi}{d\varphi^2} + m^2\Phi = 0 \tag{3.21}$$

and

$$\left[\frac{1}{\sin\theta}\frac{d}{d\theta}\left(\sin\theta\frac{d}{d\theta}\right) + \lambda - \frac{m^2}{\sin^2\theta}\right]\Theta(\theta) = 0 \tag{3.22}$$

the first is the well-known equation of harmonic motion (Atkin, 1959), whose regular solutions in complex form are

$$\Phi_m(\varphi) = \frac{1}{\sqrt{2\pi}}\exp(i\,m\varphi) \quad m = 0, \pm1, \pm2, \ldots \tag{3.23}^4$$

and the second can be written in the form

$$\frac{d^2\Theta}{d\theta^2} + \cot\theta\frac{d\Theta}{d\theta} + \left(\lambda - \frac{m^2}{\sin^2\theta}\right)\Theta = 0 \tag{3.24}$$

---

[4] m is found to be *integer* from the single valuedness requirement imposed upon the function Φ.

It is immediately evident from Equation 3.24 that the solutions Θ will depend on, besides λ, the absolute value of the quantum number $|m|$, while $\theta = 0$ is a disturbing point in the interval of definition of the variable θ ($0 \leq \theta \leq \pi$), since $\sin\theta = 0$ there and the coefficients of Θ and Θ' diverge. A point like this is known in mathematics as a *regular singularity* (Ince, 1963). Changing to the variable $\cos\theta = x(-1 \leq x \leq 1)$, the asymptotic study of the resulting differential equation in the vicinity of the singular points $x = \pm 1$

$$(1-x^2)\frac{d^2\Theta}{dx^2} - 2x\frac{d\Theta}{dx} + \left(\lambda - \frac{m^2}{1-x^2}\right)\Theta = 0 \qquad (3.25)$$

is seen to have regular solutions when

$$\Theta(x) = (1-x^2)^{m/2}G(x) \qquad (3.26)$$

where $G(x)$ is a new function to be determined and we have put $|m| = m \geq 0$ for short. Upon substitution we obtain for $G(x)$ the differential equation[5]

$$(1-x^2)\frac{d^2G}{dx^2} - 2(m+1)x\frac{dG}{dx} + [\lambda - m(m+1)]G = 0 \qquad (3.27)$$

which is again solved by the power series expansion in the variable x:

$$G(x) = \sum_k a_k x^k \qquad (3.28)$$

Proceeding in much the same way as we did for the radial equation, we obtain for the expansion coefficients the two-term recursion formula

$$a_{k+2} = \frac{(k+m)(k+m+1) - \lambda}{(k+1)(k+2)}a_k \quad k = 0, 1, 2, \ldots \qquad (3.29)$$

---

[5] Which is now free from singularities at $x = \pm 1$.

where we obtain this time an *even* and an *odd* series. Study of the convergence of the series shows that (3.28) is divergent at $|x| = 1$, so that once again the series must be truncated to a polynomial. This is possible if

$$a_k \neq 0, \quad a_{k+2} = a_{k+4} = \cdots = 0 \tag{3.30}$$

Namely, if the numerator of (3.29) vanishes, then

$$(k+m)(k+m+1) - \lambda = 0 \tag{3.31}$$

giving[6]

$$\lambda = (k+m)(k+m+1) \quad k,m = 0,1,2,\ldots \tag{3.32}$$

Posing

$$k+m = \ell \quad \text{a non-negative integer } (\ell \geq 0) \tag{3.33}$$

we obtain

$$\ell = m, \ m+1, \ m+2, \ \ldots \quad \ell \geq |m| \quad -\ell \leq m \leq \ell \tag{3.34}$$

and we recover the well-known relation between angular quantum numbers ℓ and m. Hence, we obtain for the eigenvalue of $\hat{L}^2$

$$\lambda = \ell(\ell+1) \tag{3.35}$$

$$\ell = 0,1,2,3,\ldots,(n-1) \tag{3.36}$$

$$m = 0, \ \pm 1, \ \pm 2, \ \pm 3,\ldots, \ \pm\ell \tag{3.37}$$

From (3.28) we have two polynomials, giving for the complete solution of the angular equation (3.25)

---

[6] Remember that we are using m for $|m| \geq 0$.

$$\Theta_{\ell m}(x) = (1-x^2)^{m/2} \left[\sum_{k=0}^{(\ell-m)/2} a_{2k}x^{2k} + \sum_{k=0}^{(\ell-m-1)/2} a_{2k+1}x^{2k+1} \right] \quad (3.38)$$

where the first term in brackets is the *even* polynomial and the second term is the *odd* polynomial, whose degree is at most $k_{\max} = \ell - m(\geq 0)$ for both.

Using (3.38) together with the recursion formula (3.29) with $\lambda = \ell(\ell+1)$, we obtain the following for the first few angular solutions $(x = \cos\theta)$:

$$\begin{cases}
\ell = 0, \; m = 0 & \ell - m = 0 & \Theta_{00} = a_0 \\
\ell = 1, \; m = 0 & \ell - m = 1 & \Theta_{10} = xa_1 \propto \cos\theta \\
\ell = 2, \; m = 0 & \ell - m = 2 & \Theta_{20} = a_0 + a_2x^2 = (1-3x^2)a_0 \\
\ell = 1, \; m = 1 & \ell - m = 0 & \Theta_{11} = (1-x^2)^{1/2}a_0 \propto \sin\theta \\
\ell = 2, \; m = 1 & \ell - m = 1 & \Theta_{21} = (1-x^2)^{1/2}xa_1 \propto \sin\theta\cos\theta \\
\ell = 2, \; m = 2 & \ell - m = 0 & \Theta_{22} = (1-x^2)a_0 \propto \sin^2\theta \\
\ell = 3, \; m = 2 & \ell - m = 1 & \Theta_{32} = (1-x^2)xa_1 \propto \sin^2\theta\cos\theta
\end{cases}$$

$$(3.39)$$

where a_0 and a_1 are normalization factors for the even polynomials and odd polynomials respectively. It is seen that the angular solutions (3.39) are combinations of simple trigonometric functions proportional to the associated Legendre polynomials $P_\ell^m(x)$, well known in mathematics in potential theory (Abramowitz and Stegun, 1965; Hobson, 1965) with a proportionality factor

$$P_\ell^m(x) = (-1)^{m+[(\ell+m)/2]}\Theta_{\ell m}(x) \quad (3.40)$$

where $[\cdots]$ stands for 'integer part of', a factor anyway irrelevant from the standpoint of the differential equation. The skilled reader can verify the solutions (3.39) by direct substitution in the differential equation (3.25) with $\lambda = \ell(\ell+1)$.

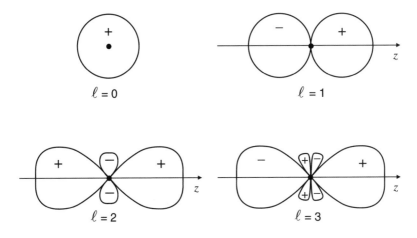

Figure 3.2 Schematic polar diagrams of the angular part of s, p, d, and f AOs with $m = 0$

The polar diagrams of the angular functions $\Theta_{\ell 0}$ with[7] $\ell = 0, 1, 2, 3$ are sketched in Figure 3.2.

3.2.4 Some Properties of the Hydrogen-like Atomic Orbitals

The HAOs in complex form

$$\psi_{n\ell m}(r, \theta, \varphi) = R_{n\ell}(r) Y_{\ell m}(\theta, \varphi) \qquad (3.41)$$

satisfy simultaneously the following three eigenvalue equations:

$$\begin{cases} \hat{H}\psi_{n\ell m} = E_n\psi_{n\ell m} & n = 1, 2, 3, \ldots \\ \hat{L}^2\psi_{n\ell m} = \ell(\ell+1)\psi_{n\ell m} & \ell = 0, 1, 2, 3, \ldots, (n-1) \\ \hat{L}_z\psi_{n\ell m} = m\psi_{n\ell m} & m = 0, \pm 1, \pm 2, \pm 3, \ldots, \pm\ell \end{cases} \qquad (3.42)$$

where $\hat{H}$ is given by (3.1) and E_n by (3.5). Each eigenvalue is character-ized by one quantum number. So, the principal quantum number n

---

[7] It is seen that ℓ equals the number of nodal planes of the functions.

characterizes the energy levels and the orbital quantum number ℓ characterizes the square of the angular momentum, the so-called magnetic number m, the z-component of the angular momentum.

Now, a fundamental theorem of quantum mechanics (Eyring *et al.*, 1944) states that, when the same function is simultaneously an eigenfunction of different operators, the corresponding operators commute in pairs; namely:

$$[\hat{H}, \hat{L}^2] = [\hat{H}, \hat{L}_z] = [\hat{L}^2, \hat{L}_z] = 0 \qquad (3.43)$$

Operators commuting with the Hamiltonian are called *constants of the motion*, which means that energy, the square of the angular momentum and the z-component of the angular momentum can all be measured with arbitrary accuracy at the same time.

The discrete spectrum of the first few energy levels of the bound hydrogen-like electron up to $n = 3$ is given in Figure 3.3. It is seen that,

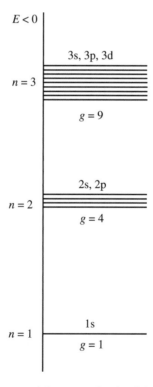

Figure 3.3 Schematic diagram of the energy levels of the hydrogen-like atom up to $n = 3$

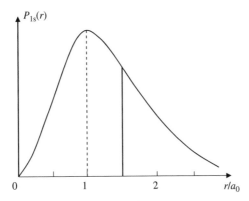

Figure 3.4   Radial probability density for the H atom ground state

apart from the 1s ground state level, which is not degenerate, all excited levels are n^2-fold degenerate. The degeneracy g of the excited energy levels is partly removed in the many-electron atom.

Lastly, we recall from first principles that

$$|\psi_{n\ell m}(r)|^2 \, d\mathbf{r} = \text{probability of finding in } d\mathbf{r} \text{ the electron in state } \psi_{n\ell m}$$
(3.44)

Integrating over angles, we obtain the probability of finding the electron inside a spherical shell of radius comprised between r and $r + dr$, the so-called *radial* probability:

$$P_{n\ell}(r) \, dr = [R_{n\ell}(r)]^2 r^2 \, dr \qquad (3.45)$$

where $P_{n\ell}(r)$ is the radial probability density.

Figure 3.4 gives the radial probability density for the 1s ground state of the H atom and the average value of the distance of the electron from the nucleus $\langle r \rangle_{1s}$. It is seen that, while the probability density $P_{1s}(r)$ has a maximum at $r = a_0$ (the Bohr radius), $\langle r \rangle_{1s} = 1.5 a_0$ (the vertical bar in the figure).

3.2.5   Real Form of the Atomic Orbitals

In valence theory, it is customary to use the *real* form for the angular part of the AOs, which is the same for all orbitals, even those that are not hydrogen-like. It is important to note that real AOs are no longer eigenfunctions of the operator $\hat{L}_z$, so that the last of the equations in Equation 3.42 is not true for real AOs. It should also be noted that in quantum mechanics the Φ-functions (3.23) in complex form are usually given with the so-called Condon–Shortley phase (Brink and Satchler, 1993):

$$\begin{cases} \Phi_{+m} = (-1)^m \dfrac{\exp(im\phi)}{\sqrt{2\pi}} & m > 0 \\[2mm] \Phi_{-m} = \dfrac{\exp(-im\phi)}{\sqrt{2\pi}} = (-1)^m (\Phi_{+m})^* \end{cases} \tag{3.46}$$

It is then possible to convert *complex* to *real* AOs using Euler's formula for imaginary exponentials:

$$\exp(\pm im\varphi) = \cos m\varphi \pm i \sin m\varphi \tag{3.47}$$

We obtain

$$\begin{cases} \Phi_m^c = \dfrac{(-1)^m \Phi_{+m} + \Phi_{-m}}{\sqrt{2}} = \dfrac{\cos m\phi}{\sqrt{\pi}} \\[2mm] \Phi_m^s = -i\dfrac{(-1)^m \Phi_{+m} - \Phi_{-m}}{\sqrt{2}} = \dfrac{\sin m\phi}{\sqrt{\pi}} \end{cases} \tag{3.48}$$

where normalization is preserved during the unitary transformation $(\mathbf{U}^{-1} = \mathbf{U}^\dagger)$. The inverse transformation (from real to complex functions) is

$$\begin{cases} \Phi_{+m} = (-1)^m \dfrac{\Phi_m^c + i\Phi_m^s}{\sqrt{2}} \\[2mm] \Phi_{-m} = \dfrac{\Phi_m^c - i\Phi_m^s}{\sqrt{2}} \end{cases} \tag{3.49}$$

Spherical harmonics in real form are called *tesseral harmonics* in mathematics (MacRobert, 1947), and in un-normalized form are written as

$$Y_{\ell 0}, \quad Y_{\ell m}^c \propto \Theta_{\ell m} \cos m\varphi, \quad Y_{\ell m}^s \propto \Theta_{\ell m} \sin m\varphi \quad (m > 0) \qquad (3.50)$$

The first few (un-normalized) HAOs in real form are hence simply given by

$$\begin{cases} 1s \propto \exp(-cr) & (c = Z) \\ 2s \propto \exp(-cr)(1 - cr) & (2c = Z) \\ 2p_z \propto \exp(-cr)z \\ 2p_x \propto \exp(-cr)x \\ 2p_y \propto \exp(-cr)y \end{cases} \qquad (3.51)$$

$$\begin{cases} 3s \propto \exp(-cr)\left(1 - 2cr + \frac{2}{3}c^2 r^2\right) & (3c = Z) \\ \\ 3p_z \propto \exp(-cr)\left(1 - \frac{1}{2}cr\right)z \\ \\ 3p_x \propto \exp(-cr)\left(1 - \frac{1}{2}cr\right)x \\ \\ 3p_y \propto \exp(-cr)\left(1 - \frac{1}{2}cr\right)y \end{cases} \qquad (3.52)$$

$$\begin{cases} 3d_{z^2} \propto \exp(-cr)\dfrac{3z^2 - r^2}{2} \\ 3d_{xy} \propto \exp(-cr)xy \\ 3d_{zx} \propto \exp(-cr)zx \\ 3d_{yz} \propto \exp(-cr)yz \\ 3d_{x^2-y^2} \propto \exp(-cr)(x^2 - y^2) \end{cases} \qquad (3.53)$$

where x, y, and z are given by Equation 1.45. Hence, our *real* AOs have the same transformation properties of (x,y,z)-coordinates or of their combinations. Replacing the radial polynomial part of the HAOs by r^{n-1}, the functional dependence of these real HAOs on (r, θ, φ) is the *same* as that of STOs with the orbital exponent c considered as a variable parameter, as we shall see below.

3.3   SLATER-TYPE ORBITALS

STOs were introduced long ago for the many-electron atoms by Slater (1930) and Zener (1930). Slater showed that for many purposes only the term with the highest power of r in the hydrogen-like $R_{n\ell}(r)$ is of importance in practical calculations. Zener suggested that replacing the hydrogenic orbital exponent $c = Z/n$ by an effective nuclear charge $(Z_{\text{eff}} = Z - s)^8$ could account in some way for the *screening* of the nuclear charge Z by the remaining electrons, giving in this way orbitals that are more diffuse than the original HAOs. Slater also gave some rules for estimating the screening constant s, which are today replaced by the variational estimation of the orbital exponents c.

Retaining only the highest $(n - \ell - 1)$ power of r of the polynomial in (3.15), the dependence on ℓ is lost and we can write the general STO in real form as

$$\chi_{n\ell m}(r, \theta, \varphi) = |n\ell m\rangle = N \exp(-cr) r^{n-1} Y_{\ell m}^{c,s}(\theta, \varphi) = R_n(r) Y_{\ell m}^{c,s}(\theta, \varphi) \tag{3.54}$$

where N is a normalization factor, R_n and $Y_{\ell m}^{c,s}$ are each individually normalized to 1, and $c > 0$ is the orbital exponent considered as a variable parameter. $Y_{\ell m}^{c,s}$ are the tesseral harmonics (3.50), which are best expressed in terms of the associated Legendre polynomials $P_\ell^m(x)$ $(x = \cos\theta)$ as

$$Y_{\ell m}^{c,s}(\theta, \varphi) = N_\Omega P_\ell^m(x) \begin{cases} \cos m\varphi \\ \sin m\varphi \end{cases} \quad m \geq 0 \tag{3.55}$$

---

[8] $s\, (< Z)$ was called by Zener the screening constant.

As we have already said, the only difference with STOs from the hydrogenic AOs (3.51)–(3.53) is in the radial part $R_n(r)$, which is now independent of ℓ. This implies a wrong behaviour of the ns functions with $n > 1$, which are now zero at the origin, lacking the characteristic cusp present in HAOs and which must be eventually corrected by Schmidt orthogonalization of the ns STO against all its inner s orbitals.

Bearing in mind the characteristic integrals (Abramowitz and Stegun, 1965)

$$\int_0^\infty dr\, r^n \exp(-cr) = \frac{n!}{c^{n+1}} \tag{3.56}$$

$$\int_0^{2\pi} d\varphi \left\{ \begin{array}{l} \cos^2 m\varphi \\[2mm] \sin^2 m\varphi \end{array} \right. = \pi \left\{ \begin{array}{l} (1+\delta_{0m}) \\[2mm] (1-\delta_{0m}) \end{array} \right. \tag{3.57}$$

$$\int_{-1}^1 dx\, [P_\ell^m(x)]^2 = \frac{2}{2\ell+1} \frac{(\ell+m)!}{(\ell-m)!} \quad m = |m| \geq 0 \tag{3.58}$$

the overall normalization factor N of the general STO is obtained as

$$N = N_r N_\Omega = \left[\frac{(2c)^{2n+1}}{(2n)!} \right]^{1/2} \left[\frac{2\ell+1}{2\pi(1+\delta_{0m})} \frac{(\ell-m)!}{(\ell+m)!} \right]^{1/2} \tag{3.59}$$

where N_r and N_Ω are separate radial and angular normalization factors. It must be further recalled that the *real* spherical harmonics (tesseral harmonics) are orthogonal and normalized to 1:

$$\int d\Omega\, Y_{\ell'm'}^{c,s}(\Omega) Y_{\ell m}^{c,s}(\Omega) = \delta_{\ell\ell'}\delta_{mm'} \tag{3.60}$$

where Ω stands for the angular variables (θ, φ) and $d\Omega = \sin\theta\, d\theta\, d\varphi$.

It is left as an exercise for the reader to verify that the first few normalized STOs are

$$1s = \left(\frac{c_0^3}{\pi}\right)^{1/2} \exp(-c_0 r) \tag{3.61}$$

$$s = \left(\frac{c_s^5}{3\pi}\right)^{1/2} \exp(-c_s r) r \tag{3.62}$$

$$2p_z = \left(\frac{c_p^5}{\pi}\right)^{1/2} \exp(-c_p r) r \cos\theta \tag{3.63}$$

with the matrix elements of the hydrogenic Hamiltonian being

$$\left\langle 1s \left| -\frac{1}{2}\nabla^2 - \frac{Z}{r} \right| 1s \right\rangle = \frac{c_0^2}{2} - Z c_0 \tag{3.64}$$

$$\left\langle s \left| -\frac{1}{2}\nabla^2 - \frac{Z}{r} \right| s \right\rangle = \frac{c_s^2}{6} - \frac{1}{2} Z c_s \tag{3.65}$$

$$\left\langle 2p_z \left| -\frac{1}{2}\nabla^2 - \frac{Z}{r} \right| 2p_z \right\rangle = \frac{1}{2}(c_p^2 - Z c_p) \tag{3.66}$$

$$\left\langle s \left| -\frac{1}{2}\nabla^2 - \frac{Z}{r} \right| 1s \right\rangle = S_{s1s}\left[-\frac{c_0^2}{2} + (c_0 - Z)\frac{c_0 + c_s}{3} \right] \tag{3.67}$$

where S_{s1s} is the nonorthogonality integral between 1s and s STOs:

$$S_{s1s} = \langle s|1s\rangle = \sqrt{3}\,\frac{c_s}{c_0 + c_s}\left[\frac{2(c_0 c_s)^{1/2}}{c_0 + c_s}\right]^3 \tag{3.68}$$

These formulae will be needed in the Chapter 4.

3.4 GAUSSIAN-TYPE ORBITALS

Gaussian orbitals are largely used today in atomic and molecular computations because of their greater simplicity in computing multicentre molecular integrals. GTOs were introduced by Boys (1950) and McWeeny (1950) mostly for computational reasons, but they are definitely inferior to STOs because of their incorrect radial behaviour, both at the origin[9] and in the tail of the wavefunction. In today's molecular calculations it is customary to fit STOs in terms of GTOs, which requires rather lengthy expansions with a large number of terms. Some failures of GTOs with respect to STOs are briefly discussed in the Chapter 4.

Most common Gaussian orbitals are given in the form of spherical or Cartesian functions. We shall give some definitions here and a few integrals which are needed later.

3.4.1 Spherical Gaussians

For spherical GTOs it will be sufficient to consider only the radial part of the orbital, since the angular part is the same as that for STOs:

$$R_n(r) = N_n \exp(-cr^2) r^{n-1} \quad N_n = \left[\frac{2^{n+1}(2c)^{n+(1/2)}}{(2n-1)!!\sqrt{\pi}} \right]^{1/2} \quad (c > 0)$$

$$(3.69)$$

where

$$(2n-1)!! = (2n-1)(2n-3)\ldots 3 \times 1 = \frac{(2n)!}{2^n n!} = \frac{(2n)!}{(2n)!!} \quad (-1)!! = 0!! = 1$$

$$(3.70)$$

is the double factorial.

This normalization factor in (3.69) can be derived from the general Gaussian integral

$$\int_0^\infty dr \exp(-cr^2) r^n = \frac{(n-1)!!}{(2c)^{(n+1)/2}} \sigma(n)$$

$$(3.71)$$

---

[9] Spherical GTOs lack the characteristic cusp at the origin.

where

$$\sigma(n) = \sqrt{\frac{\pi}{2}} \text{ for } n = \text{even}, \quad \sigma(n) = 1 \text{ for } n = \text{odd} \qquad (3.72)$$

The normalized spherical Gaussian orbital will then be written as

$$|n\ell m\rangle = G(n\ell m, c) = N \exp(-cr^2)r^{n-1}Y_{\ell m}^{c,s}(\theta, \varphi) \qquad (3.73)$$

with $N = N_n N_\Omega$.

A few results for 1s GTOs are as follows:

$$\left\langle G(100, c') \left| -\frac{1}{2}\nabla^2 - \frac{Z}{r} \right| G(100, c) \right\rangle = \sqrt{\frac{72(cc')^{7/2}}{(c+c')^5}} - Z\sqrt{\frac{32(cc')^{3/2}}{(c+c')^2\pi}}$$

$$(3.74)$$

and, if $c' = c$:

$$\left\langle G(100, c) \left| -\frac{1}{2}\nabla^2 - \frac{Z}{r} \right| G(100, c) \right\rangle = \frac{3}{2}c - Z\left(\frac{8c}{\pi}\right)^{1/2} \qquad (3.75)$$

It is apparent that these formulae are sensibly less appealing than those found for STOs.

3.4.2 Cartesian Gaussians

The normalized Cartesian Gaussians are best written as

$$G_{uvw}(c) = N \exp(-cr^2)x^u y^v z^w = N \exp[-c(x^2 + y^2 + z^2)]x^u y^v z^w$$

$$(3.76)$$

where u, v, and w are non-negative integers and the normalization integral is given by

$$N = \left[\frac{(4c)^{U+V+W}}{(2U-1)!!(2V-1)!!(2W-1)!!} \left(\frac{2c}{\pi} \right)^{3/2} \right]^{1/2} \qquad (3.77)$$

where

$$2U = u + u', \quad 2V = v + v', \quad 2W = w + w' \qquad (3.78)$$

4

The Variation Method

In this chapter we present elements of the variation method, the most powerful technique for doing working approximations when the Schroedinger eigenvalue equation cannot be solved exactly. Applications involve optimization of variational parameters, either nonlinear (orbital exponents) or linear (the Ritz method). To illustrate the method, examples on ground and excited states of the H-like and the He-like atomic systems are worked out in some detail.

4.1 VARIATIONAL PRINCIPLES

Let φ be a normalizable regular *trial* (or variational) function. We define the Rayleigh ratio as the functional[1]

$$\varepsilon[\varphi] = \frac{\langle \varphi | \hat{H} | \varphi \rangle}{\langle \varphi | \varphi \rangle} = \frac{\int dx\, \varphi^*(x) \hat{H} \varphi(x)}{\int dx\, \varphi^*(x) \varphi(x)} \qquad (4.1)$$

where x are the electronic coordinates and $\hat{H}$ the Hamiltonian of the system. Then

$$\varepsilon[\varphi] \geq E_0 \qquad (4.2)$$

---

[1] A function of function $\varphi(x)$.

Methods of Molecular Quantum Mechanics: An Introduction to Electronic Molecular Structure
Valerio Magnasco
© 2009 John Wiley & Sons, Ltd

is the Rayleigh variational principle for the ground state, E_0 being the *true* ground state energy;

$$\varepsilon[\varphi] \geq E_1 \quad \text{provided} \quad \langle \psi_0 | \varphi \rangle = 0 \tag{4.3}$$

is the Rayleigh variational principle for the first excited state, provided the trial function φ is taken *orthogonal* to the *true* ground state function ψ_0. The proofs of these statements are given in Magnasco (2007).

Therefore, evaluation of the integrals in (4.1) under the suitable constraints of normalization and orthogonality gives the *upper bounds* to the energy of the ground and excited states depicted in Figure 4.1. This is of fundamental importance in applications, since the variational energy must always lie *above* the true energy.

It is easily seen that:

- The equality sign holds for the exact functions.
- If the variational function is affected by a first-order error, then the error in the variational energy is second order. Therefore, energy is

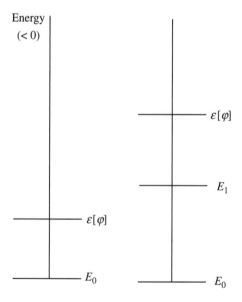

Figure 4.1 Energy upper bounds to ground (left) and first excited state (right)

always determined better than the wavefunction. The same is true for the second-order energies of Chapter 10.

- The variational method privileges the regions of space *near* the nucleus, so that variationally determined wavefunctions may be not appropriate for dealing with the expectation values of operators that take large values *far* from the nucleus, like the electric dipole moment.

Variational approximations to energy and the wavefunction can be worked out simply by introducing some *variational parameters* $\{c\}$ in the trial function and then evaluating the integrals in the functional (4.1), giving in this way an ordinary function of the variational parameters $\{c\}$, which has to be *minimized* against the parameters. For a single parameter c (Figure 4.2)

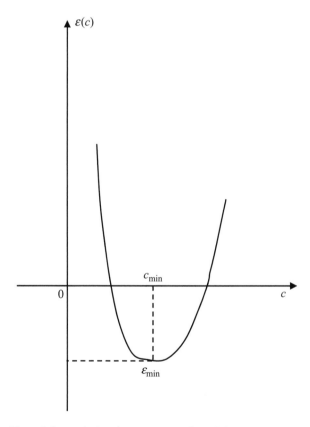

Figure 4.2   Plot of the variational energy near the minimum versus c

$$\varepsilon[\varphi] = \frac{\int dx\, \varphi^*(x;c)\hat{H}\varphi(x;c)}{\int dx\, \varphi^*(x;c)\varphi(x;c)} = \varepsilon(c) \qquad (4.4)$$

$$\frac{d\varepsilon}{dc} = 0 \Rightarrow c_{\min} \qquad (4.5)$$

provided

$$\left(\frac{d^2\varepsilon}{dc^2}\right)_{c_{\min}} > 0 \qquad (4.6)$$

In this way, we obtain the *best* approximation compatible with the form assumed for the approximate trial function. Increasing the number of flexible parameters increases the accuracy of the variational result.[2]

For N variational parameters, $\{c\} = (c_1, c_2, \ldots, c_N)$, Equations 4.4 and 4.5 must be replaced by

$$\varepsilon[\varphi] = \varepsilon(c_1, c_2, \ldots, c_N) \qquad (4.7)$$

$$\frac{\partial \varepsilon}{\partial c_1} = \frac{\partial \varepsilon}{\partial c_2} = \cdots = \frac{\partial \varepsilon}{\partial c_N} = 0 \Rightarrow \{c(\text{best})\} \qquad (4.8)$$

$$\varepsilon(\text{best}) = \varepsilon(c_1^0, c_2^0, \ldots, c_N^0) \qquad (4.9)$$

$$\varphi(\text{best}) = \varphi(x; c_1^0, c_2^0, \ldots, c_N^0) \qquad (4.10)$$

where $\{c(\text{best})\} = (c_1^0, c_2^0, \ldots, c_N^0)$ is the set of N optimized parameters.

For working approximations, we must resort to some basis set of regular functions (such as the STOs or GTOs of Chapter 3), introducing either (i) nonlinear (orbital exponents) or (ii) linear variational parameters.

---

[2] Using appropriate numerical methods, it is feasible today to optimize variational wavefunctions containing millions of terms, such as those encountered in the configuration interaction techniques of Chapter 8.

4.2   NONLINEAR PARAMETERS

In this case, we cannot obtain any standard equations for the optimization, which must usually be done by numerical methods (e.g. see the
simple Ransil method, useful for functions having a parabolic behaviour
near the minimum). The method is rather powerful in itself, but, when
many parameters are involved, there may be some troubles with the
difficulty of avoiding spurious secondary minima in the energy hypersur-
face, which would spoil the numerical results.

In the following, we shall concisely apply this method to simple
variational functions containing a *single* variational parameter c (>0)
for the one-electron and the two-electron atomic cases. The H-like system
is particularly important, mostly for didactical reasons, since we
can compare our approximate results with the exact ones derived in
Chapter 3.

4.2.1   Ground State of the Hydrogenic System

We compare the relative accuracy of three variational functions, two of
STO type and one GTO, each containing a single adjustable orbital
exponent c for the ground state of the hydrogen-like system having the
one-electron Hamiltonian (in atomic units):

$$\hat{h} = -\frac{1}{2}\nabla^2 - \frac{Z}{r} \tag{4.11}$$

where Z is the nuclear charge ($Z = 1$ for hydrogen). The expressions of
the energy integral and those of the analytical optimization[3] are collected
in Table 4.1 and the numerical results are given in Table 4.2.

All functions in the first column of both tables are un-normalized, and
must be normalized at the end of each calculation. φ_1 is the single-
parameter trial function having the correct functional form as the exact
wavefunction ψ_0. Optimization of the nonlinear variational parameter c
yields the correct wavefunction and energy, corresponding to the equality sign in (4.2). φ_2 is a Slater 2s, which lacks the cusp characteristic of the
true ns hydrogenic wavefunctions (it is zero at the origin) and decreases
too slowly in its tail because of r in its functional form. It gives the worst

---

[3] The students are strongly recommended to do the calculations by themselves.

Table 4.1 Variational approximations to the ground state of the hydrogenic system

φ	$\varepsilon(c)$	c_{min}	ε_{min}
$\varphi_1 = \exp(-cr)$	$\dfrac{c^2}{2} - Zc$	Z	$-\dfrac{Z^2}{2}$
$\varphi_2 = \exp(-cr)r$	$\dfrac{c^2}{6} - \dfrac{Zc}{2}$	$\dfrac{3}{2}Z$	$-\dfrac{3}{8}Z^2$
$\varphi_3 = \exp(-cr^2)$	$\dfrac{3}{2}c - Z\left(\dfrac{8c}{\pi}\right)^{1/2}$	$\dfrac{8}{9\pi}Z^2$	$-\dfrac{4}{3\pi}Z^2$

Table 4.2 Numerical results of different variational approximations for the ground-state H atom

φ	c_{min}	$\varepsilon(c_{min})/E_h$
$\varphi_1 = \exp(-cr)$	1	-0.5
$\varphi_2 = \exp(-cr)r$	1.5	-0.375
$\varphi_3 = \exp(-cr^2)$	0.2829	-0.4244

energy result, with no more than 75% of the truth. The third function φ_3 is the prototype of the 1s Gaussian function. Even if it has the wrong behaviour at the origin (it has a zero derivative here) and decreases too quickly far from the nucleus, it nevertheless gives a fair result for the energy, namely about 85% of the true value, which is 10% better than φ_2. However, increasing the number of 1s GTOs improves the energy, but does not improve the wavefunction sufficiently; even taking $N = 10$ optimized 1s GTOs, it is still very different from the correct one.[4] It is interesting to note from Table 4.2 that the variational theorem tries to do its best to correct the inappropriate form of functions φ_2 and φ_3 by strongly increasing ($c_{min} = 1.5$, the function *contracts*) or decreasing ($c_{min} = 0.2829$, the function *expands*) respectively the best value for their orbital exponents. Taking $c = 1$ in φ_2 would give $\varepsilon_\varphi = -0.333\,333E_h$, which is about 67% of the true value. Hence, optimization of c improves energy by 8%.

---

[4] The incorrect behaviour at the origin of 1s GTOs and the way of correcting for it are fully discussed by Magnasco (2007).

4.2.2 The First Excited State of Spherical Symmetry of the Hydrogenic System

The first spherical excited state of the H atom is the 2s state, whose energy belongs to the four degenerate levels with $n = 2$. As with all ns orbitals, the correct function has a cusp at the origin. The spherical normalized STO

$$s = \left(\frac{c_s^5}{3\pi}\right)^{1/2} \exp(-c_s r)r \qquad (4.12)$$

is not orthogonal to the ground-state 1s function (orbital exponent $c_0 = 1$), with the nonorthogonality integral

$$S = \langle \psi_0 | s \rangle = \sqrt{3}\frac{c_s}{c_s + 1}\left(\frac{2\sqrt{c_s}}{c_s + 1}\right)^3 \qquad (4.13)$$

and cannot be used as such in the variational calculation of the energy of the excited state. In fact, without the constraint of orthogonality demanded by (4.3), function s would give at most, as we have just seen, a poor bound to the ground-state energy.

A convenient variational function for the first excited state of spherical symmetry is obtained by first orthogonalizing s against ψ_0 by the Schmidt method, giving the normalized trial function

$$\varphi = \frac{s - S\psi_0}{\sqrt{1 - S^2}} \qquad \langle \psi_0 | \varphi \rangle = 0 \qquad (4.14)$$

which is now orthogonal to ψ_0, as it must be. It is then easily seen that the new variational function (4.14) gives the upper bound

$$\varepsilon_\varphi = \left\langle \varphi \left| -\frac{1}{2}\nabla^2 - \frac{Z}{r} \right| \varphi \right\rangle = h_{ss} + E_{pn} \geq E_{2s} \qquad (4.15)$$

where

$$h_{ss} = \frac{c_s^2}{6} - \frac{Zc_s}{2} \qquad (4.16)$$

is the energy of the nonorthogonal function (4.12) and

$$E_{pn} = \frac{S^2 (h_{\psi_0 \psi_0} + h_{ss}) - 2S\, h_{s\psi_0}}{1 - S^2} = -S \frac{(\psi_0 s - S\psi_0^2 |\hat{h}) + (s\psi_0 - Ss^2|\hat{h})}{1 - S^2} > 0$$

(4.17)

the term correcting for nonorthogonality. This *repulsive* term, called the penetration (or exchange-overlap) energy, avoids the excited 2s electron from collapsing onto the inner 1s electron. It is easily seen by direct calculation that the nondiagonal matrix element in (4.17) is given by (compare with Equation 3.67)

$$h_{s\psi_0} = \left\langle s \left| -\frac{1}{2} \nabla^2 - \frac{Z}{r} \right| \psi_0 \right\rangle = \frac{S}{3} \left[(c_0 - Z)(c_0 + c_s) - \frac{3}{2} c_0^2 \right]$$

(4.18)

which, for $c_0 = Z = 1$, reduces to

$$h_{s\psi_0} = -\frac{S}{2}$$

(4.19)

The behaviour of the three different components of the variational energy calculation for $Z = 1$ are plotted qualitatively in Figure 4.3 as a

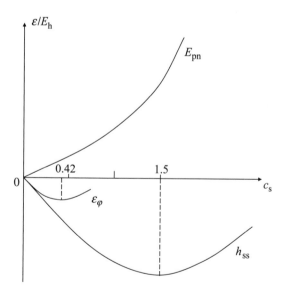

Figure 4.3 Plots of the variational energy components for H(2s) versus the c_s orbital exponent

function of the orbital exponent c_s. The too attractive h_{ss} has its bad minimum at $c_s = 1.5$ (see Table 4.2) and is contrasted by the repulsive E_{pn} so as to give the *best* variational minimum of $\varepsilon_\varphi = -0.1234E_h$ occurring for $c_s = 0.4222$. This result is within 98.7% of the exact result, $E_{2s} = -0.125E_h$. Even without doing orbital exponent optimization $(c_s = 0.5)$, the remarkable result of $\varepsilon_\varphi = -0.1192E_h$ would be obtained (about 95% of the exact value), showing the crucial importance of Schmidt's orthogonalization in assessing a reasonable variational approximation to the 2s excited state.

4.2.3 The First Excited 2p State of the Hydrogenic System

We can take as an appropriate variational function for this case the normalized $2p_z$ STO:

$$\varphi = \left(\frac{c_p^5}{\pi}\right)^{1/2} \exp(-c_p r) r \cos\theta \tag{4.20}$$

orthogonal to ψ_0 by symmetry. The orthogonality constraint is now satisfied from the outset, and simple calculation gives the upper bound (Equation 3.66):

$$\varepsilon_\varphi = \left\langle 2p_z \left| -\frac{1}{2}\nabla^2 - \frac{Z}{r} \right| 2p_z \right\rangle = \frac{1}{2}(c_p^2 - Zc_p) \geq E_{2p} \tag{4.21}$$

giving upon c_p optimization the *exact* value for the excited 2p energy level of the H-like system:

$$\frac{d\varepsilon_\varphi}{dc_p} = c_p - \frac{Z}{2} = 0 \Rightarrow c_p(\text{best}) = \frac{Z}{2} \Rightarrow \varepsilon_\varphi(\text{best}) = -\frac{Z^2}{8} \tag{4.22}$$

4.2.4 The Ground State of the He-like System

With reference to Figure 4.4, the two-electron Hamiltonian (in atomic units) for the He-like system is

$$\hat{H} = \hat{h}_1 + \hat{h}_2 + \frac{1}{r_{12}}, \quad \hat{h} = -\frac{1}{2}\nabla^2 - \frac{Z}{r} \tag{4.23}$$

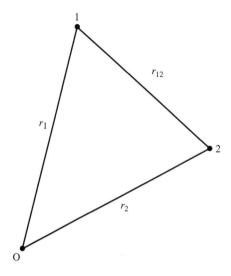

Figure 4.4   The atomic two-electron system

the sum of the two one-electron hydrogenic Hamiltonians and the electron repulsion term (a two-electron operator). This last term hinders separation of the two-electron Schroedinger eigenvalue equation into two separate equations, one for each hydrogenic electron. However, approximations are possible through the variation method using the simple two-electron product variational function:

$$\varphi(1,2) = \varphi_1(1)\varphi_2(2) \tag{4.24}$$

where the normalized one-electron functions are given by

$$\varphi_{1s} = \left(\frac{c_0^3}{\pi}\right)^{1/2} \exp(-c_0 r) \tag{4.25}$$

Calculation then gives the variational bound

$$\varepsilon_\varphi = \left\langle \varphi_1\varphi_2 \middle| \hat{h}_1 + \hat{h}_2 + \frac{1}{r_{12}} \middle| \varphi_1\varphi_2 \right\rangle = 2h_{1s1s} + (1s^2|1s^2) \tag{4.26}$$

namely:

$$\varepsilon_\varphi = 2\left(\frac{c_0^2}{2} - Zc_0\right) + \frac{5}{8}c_0 = c_0^2 - 2\left(Z - \frac{5}{16}\right)c_0 \qquad (4.27)$$

In (4.26), the first term on the right is the one-electron matrix element already found for H, while the second is the two-electron repulsion integral between the two spherical 1s charge distributions written in charge density notation:

$$\langle 1s_1\, 1s_2 | r_{12}^{-1} | 1s_1\, 1s_2 \rangle = \iint d\mathbf{r}_1\, d\mathbf{r}_2\, \frac{[1s(\mathbf{r}_2)1s^*(\mathbf{r}_2)]}{r_{12}}[1s(\mathbf{r}_1)1s^*(\mathbf{r}_1)] = (1s^2|1s^2)$$

$$(4.28)$$

This new two-electron integral is evaluated in terms of the purely *radial* electrostatic potential $J_{1s}(r_1)$:

$$J_{1s}(r_1) = \int d\mathbf{r}_2\, \frac{[1s(\mathbf{r}_2)1s^*(\mathbf{r}_2)]}{r_{12}} = \frac{1}{r_1} - \frac{\exp(-2c_0 r_1)}{r_1}(1 + c_0 r_1) \quad (4.29)$$

when use is made of the one-centre Neumann expansion for $1/r_{12}$ (Magnasco, 2007). Then, integral (4.28) is readily obtained by integration in spherical coordinates:

$$(1s^2|1s^2) = \int d\mathbf{r}_1\, J_{1s}(r_1)[1s(\mathbf{r}_1)1s^*(\mathbf{r}_1)] = \frac{5}{8}c_0 \qquad (4.30)$$

Optimization of (4.27) with respect to c_0 gives

$$c_0 = Z - \frac{5}{16} \qquad (4.31)$$

as best value for the orbital exponent and

$$\varepsilon_\varphi(\text{best}) = -\left(Z - \frac{5}{16}\right)^2. \qquad (4.32)$$

as best energy for the S($1s^2$) configuration of the He-like atom.

For $Z = 2$ (He atom), $Z_{eff} = Z - (5/16) = 27/16 = 1.6875$, and (4.32) gives $\varepsilon_\varphi(\text{best}) = -2.847\,656E_h$, which is more than 98% of the accurate value of $-2.903\,724E_h$ given by Pekeris (1958). Eckart (1930) gave as *two* nonlinear parameter approximations to the ground-state energy of the He atom the following improved 'split-shell' result:

$$\varepsilon = -2.875\,661E_h \quad c_1 = 2.183\,171 \quad c_2 = 1.188\,531 \qquad (4.33)$$

which is within 99% of Pekeris's result. We end this section by noting: (i) that Equation 4.32 describes in part the 'screening effect' of the second electron on the nuclear charge Z,[5] showing that the variation theorem accounts as far as possible for real physical effects; (ii) that the calculated energy is *not* an observable quantity, so that comparison with experimental results is possible only through the values of the *ionization potential*, defined as

$$I = \varepsilon(\text{He}^+) - \varepsilon(\text{He}) \qquad (4.34)$$

Since the ionization potential I is smaller than the absolute energies of either the atom or the ion, its approximate values are affected by larger errors.

4.3   LINEAR PARAMETERS AND THE RITZ METHOD

This famous method of linear combinations is due to the young Swiss mathematician Ritz (1909) and, therefore, is usually referred to as the Ritz method. From the variational point of view, flexibility in the trial function is now introduced through the coefficients of the linear combination of a given set of regular functions. Usually, the basis functions are *fixed*, but they can be successively optimized even with respect to the nonlinear parameters present in their functional form. We shall see that the Ritz

---

[5] In the He atom, an electron *near* to the nucleus sees the whole nuclear charge Z; *far* from it, the nuclear charge is $(Z - 1)$, as if it were fully screened by the other electron. The variational result averages between these two extreme cases, privileging the regions near to the nucleus (hence, $Z_{eff} \approx 1.7$, closer to 2 rather than 1).

method is intimately connected with the problem of the matrix diagonalization of Chapter 2.

We shall limit ourselves to consideration of a finite basis set of N
orthonormal functions χ, the problem being best treated in matrix form.

The Rayleigh ratio (4.1) can be written as

$$\varepsilon = HM^{-1} \tag{4.35}$$

where

$$H = \langle \varphi | \hat{H} | \varphi \rangle, \quad M = \langle \varphi | \varphi \rangle \tag{4.36}$$

If we introduce the set of N orthonormal functions as the *row*
matrix

$$\chi = (\chi_1 \chi_2 \cdots \chi_N) \tag{4.37}$$

and the corresponding set of variational coefficients as the *column*
matrix

$$\mathbf{c} = \begin{pmatrix} c_1 \\ c_2 \\ \cdots \\ c_N \end{pmatrix} \tag{4.38}$$

then H and M in (4.36) can be written in terms of the $(N \times N)$ Hermitian
matrices $\mathbf{H}$ and $\mathbf{M}$:

$$H = \mathbf{c}^\dagger \chi^\dagger \hat{H} \chi \mathbf{c} = \mathbf{c}^\dagger \mathbf{H} \mathbf{c}, \quad M = \mathbf{c}^\dagger \chi^\dagger \chi \mathbf{c} = \mathbf{c}^\dagger \mathbf{M} \mathbf{c} = \mathbf{c}^\dagger \mathbf{1} \mathbf{c} \tag{4.39}$$

where $\mathbf{M} = \mathbf{1}$ is the metric matrix of the basis functions χ. The matrix
elements of matrices $\mathbf{H}$ and $\mathbf{M}$ are

$$H_{\mu\nu} = \langle \chi_\mu | \hat{H} | \chi_\nu \rangle, \quad M_{\mu\nu} = \langle \chi_\mu | \chi_\nu \rangle = 1_{\mu\nu} = \delta_{\mu\nu} \tag{4.40}$$

An infinitesimal *first* variation in the linear coefficients will induce an
infinitesimal change in the energy functional (4.35):

$$\delta\varepsilon = \delta H \cdot M^{-1} - H \cdot M^{-2} \delta M = M^{-1}(\delta H - \varepsilon \delta M) \tag{4.41}$$

where to first order in $\delta \mathbf{c}$ (a column of infinitesimal variation of coefficients)

$$\delta H = \delta \mathbf{c}^\dagger \mathbf{H} \mathbf{c} + \mathbf{c}^\dagger \mathbf{H} \delta \mathbf{c}, \quad \delta M = \delta \mathbf{c}^\dagger \mathbf{1} \mathbf{c} + \mathbf{c}^\dagger \mathbf{1} \delta \mathbf{c} \qquad (4.42)$$

The necessary condition for ε being *stationary* against arbitrary variations in the coefficients yields the equation

$$\delta \varepsilon = 0 \Rightarrow \delta H - \varepsilon \delta M = 0 \qquad (4.43)$$

and in matrix form

$$\delta \mathbf{c}^\dagger (\mathbf{H} - \varepsilon \mathbf{1}) \mathbf{c} + \mathbf{c}^\dagger (\mathbf{H} - \varepsilon \mathbf{1}) \delta \mathbf{c} = 0 \qquad (4.44)$$

Because matrix $\mathbf{H}$ is Hermitian, the second term in (4.44) is simply the complex conjugate of the first, so that, since $\delta \mathbf{c}^\dagger$ is *arbitrary*, the condition (4.43) takes the matrix form

$$(\mathbf{H} - \varepsilon \mathbf{1}) \mathbf{c} = 0 \Rightarrow \mathbf{H} \mathbf{c} = \varepsilon \mathbf{c} \qquad (4.45)$$

which is the eigenvalue equation for matrix $\mathbf{H}$ (Equation 2.27). The variational determination of the linear coefficients under the constraint of orthonormality of the basis functions in the Ritz method is, therefore, completely equivalent to the problem of diagonalizing matrix $\mathbf{H}$. Following what was said there, the homogeneous system (4.45) has nontrivial solutions if and only if

$$|\mathbf{H} - \varepsilon \mathbf{1}| = 0 \qquad (4.46)$$

The solution of the secular Equation 4.46 yields as *best* values for the variational energy (4.35) the N real roots, which are usually ordered in ascending order:

$$\varepsilon_1 \leq \varepsilon_2 \leq \cdots \leq \varepsilon_N \qquad (4.47)$$

$$\mathbf{c}_1, \mathbf{c}_2, \ldots, \mathbf{c}_N \qquad (4.48)$$

$$\varphi_1, \varphi_2, \ldots, \varphi_N \qquad (4.49)$$

which are respectively the eigenvalues and eigenvectors of matrix $\mathbf{H}$ and the *best* variational approximation to the eigenfunctions. The Ritz method not only gives the best variational approximation to the ground-state energy (the first eigenvalue ε_1), but also approximations to the energy of the excited states. A theorem due to MacDonald (1933) states further that each of the ordered roots (4.47) gives an upper bound to the energy of the respective excited state.

4.4 APPLICATIONS OF THE RITZ METHOD

An application to the first two excited states of the He-like atom concludes this chapter.

4.4.1 The First 1s2s Excited State of the He-like Atom

We take as the orthonormal two-electron basis set the simple products of one-electron functions

$$\chi_1 = 1s_1\, 2s_2, \quad \chi_2 = 2s_1\, 1s_2 \tag{4.50}$$

which are assumed individually normalized and orthogonal.[6] The (2×2) secular equation is

$$\begin{vmatrix} H_{11} - \varepsilon & H_{12} \\ H_{12} & H_{22} - \varepsilon \end{vmatrix} = 0 \tag{4.51}$$

with the matrix elements

$$H_{11} = \left\langle 1s_1\, 2s_2 \left| \hat{h}_1 + \hat{h}_2 + \frac{1}{r_{12}} \right| 1s_1\, 2s_2 \right\rangle \tag{4.52}$$
$$= h_{1s1s} + h_{2s2s} + (1s^2|2s^2) = E_0 + J = H_{22}$$

$$H_{12} = \left\langle 1s_1\, 2s_2 \left| \hat{h}_1 + \hat{h}_2 + \frac{1}{r_{12}} \right| 2s_1\, 1s_2 \right\rangle = (1s\, 2s|1s\, 2s) = K \tag{4.53}$$

---

[6] The appropriate 2s AO has the same form as φ of Equation 4.14.

Both two-electron integrals are written here in charge density nota-tion, J being called the *Coulomb* integral and K (J) the *exchange* integral.

The roots are

$$\varepsilon_+ = H_{11} + H_{12} = E_0 + J + K \qquad (4.54)$$

$$\varepsilon_- = H_{11} - H_{12} = E_0 + J - K \qquad (4.55)$$

whereas the corresponding wavefunctions have definite symmetry proper-ties with respect to electron interchange:

$$\varphi_+ = \frac{1}{\sqrt{2}}(1s\,2s + 2s\,1s)$$

$$\varphi_- = \frac{1}{\sqrt{2}}(1s\,2s - 2s\,1s) \qquad (4.57)$$

the first being symmetric, the second antisymmetric in the electron interchange. The schematic diagram of the energy levels for the S(1s2s) excited state of the He-like atom is qualitatively depicted in Figure 4.5. The *splitting* between the two levels (degenerate in the absence of electron repulsion) is just $2K$; that is, twice the value of the exchange integral K. Variational optimization of the orbital exponent of the 2s function (Magnasco, 2007) for He ($Z = 2$) gives $c_s = 0.4822$, which, used in

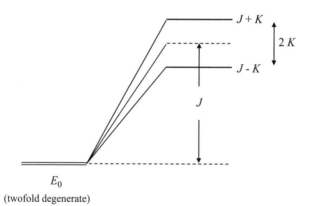

Figure 4.5 Schematic diagram of the energy levels for the excited S(1s2s) state of the He-like atom

conjunction with $c_0 = 1.6875$ for the 1s AO, yields the variational energy bounds ($E_0 = -2.348\,01E_h$, $J = 0.240\,29E_h$):

$$\varepsilon_+ = -2.093\,74E_h \quad \varepsilon_- = -2.121\,68E_h \qquad (4.58)$$

Either the excitation energies from the He ground state

$$\Delta\varepsilon_+ = 0.7539\,E_h \quad \Delta\varepsilon_- = 0.7260\,E_h \qquad (4.59)$$

or the splitting

$$2K = 0.0279E_h \qquad (4.60)$$

are seen to be in sufficiently good agreement with the experimental values (Moore, 1971) of 0.7560, 0.7282, and 0.0292 respectively. It is worth noting that hydrogenic AOs (with $c_s = 1$) would give a wrong order for the energy levels (Magnasco, 2007).

4.4.2   The First 1s2p State of the He-like Atom

The same considerations can be made for the first 1s2p excited state. Because of space degeneracy, in the absence of the electron interaction, there are now six states having the same energy, and precisely

$$\chi_1 = 1s2p_z, \chi_2 = 2p_z1s, \chi_3 = 1s2p_x, \chi_4 = 2p_x1s, \chi_5 = 1s2p_y, \chi_6 = 2p_y1s$$
$$(4.61)$$

where we have omitted for brevity the electron labels (electrons always arranged in *dictionary* order). We omit the details of the calculation, which follows strictly those already seen for the 1s2s state. We only say that functions belonging to different (x,y,z) symmetries are orthogonal and not interacting with each other and with respect to all S states. The (6×6) secular equation factorizes into three equivalent (2×2) blocks, whose matrix elements are obtained from the previous ones simply by replacing 2s by 2p, E_0 by E'_0, and J, K by J', K'. The two roots are still *triply* degenerate, so that the sixfold degeneracy occurring in the one-electron approximation E'_0 is not completely removed by consideration of the electron repulsion. Complete removal of the residual degeneracy

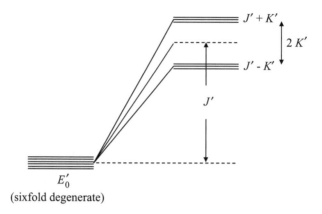

E'_0
(sixfold degenerate)

Figure 4.6 Schematic diagram of the energy levels for the excited P(1s2p) state of the He-like atom

of the energy levels is accomplished only by taking into account spin and the Zeeman effect in the presence of a magnetic field (Chapter 5). The schematic diagram of the splitting of the energy levels for the P(1s2p) state of the He-like atom is given in Figure 4.6. The evaluation in spherical coordinates of the integrals J, K, J', and K' is sketched in the Appendix.

Variational optimization of the orbital exponent of the 2p functions (Magnasco, 2007) for He $(Z = 2)$ gives $c_p = 0.4761$, which, used in conjunction with $c_0 = 1.6875$ for the 1s AO, yields the following variational energy bounds $(E'_0 = -2.313\,99E_h, \, J' = 0.236\,67E_h)$:

$$\varepsilon_+ = -2.072\,37E_h \quad \varepsilon_- = -2.082\,26E_h \qquad (4.62)$$

In this case, too, either the excitation energies from the He ground state

$$\Delta\varepsilon_+ = 0.7753\,E_h \quad \Delta\varepsilon_- = 0.7654\,E_h \qquad (4.63)$$

or the splitting

$$2K = 0.0099E_h \qquad (4.64)$$

are in reasonably good agreement with the experimental values (Moore, 1971) of 0.7796, 0.7703, and 0.0093 respectively. However, from a quantitative point of view, it is seen that the error in the variational calculation of the excited P state is *larger* than that resulting for the excited S state. The splitting between the P levels is about one order of magnitude smaller than that observed for the S levels.

APPENDIX: THE INTEGRALS *J*, *K*, *J*′ AND *K*′

The two-electron one-centre integrals J, K, J', and K' occurring in the calculation of the first excited states of He can be evaluated as one-electron integrals in spherical coordinates once the appropriate electrostatic potentials are known. With reference to the 1s, s, and $2p_z$ STOs defined by Equations 3.61–3.63, the electrostatic potentials are evaluated using the one-centre Neumann expansion for $1/r_{12}$, giving

$$J_{1s^2}(r_1) = \int dr_2 \frac{[1s(r_2)]^2}{r_{12}} = \frac{1}{r_1}[1 - \exp(-2c_0)(1 + c_0 r_1)] \tag{4.65}$$

$$J_{s1s}(r_1) = \int dr_2 \frac{[s(r_2)1s(r_2)]}{r_{12}}$$

$$= \frac{12}{(c_0 + c_s)^3} \left(\frac{c_0^3 c_s^5}{3}\right)^{1/2} \frac{1}{\rho}\left[1 - \exp(-2\rho)\left(1 + \frac{4}{3}\rho + \frac{2}{3}\rho^2\right)\right] \tag{4.66}$$

$$J_{1s2p_z}(r_1, \theta) = \int dr_2 \frac{[1s(r_2)2p_z(r_2)]}{r_{12}}$$

$$= \frac{8c_0^{3/2} c_p^{5/2}}{(c_0 + c_p)^3} \cos\theta \frac{1}{\rho^2}[1 - \exp(-2\rho)(1 + 2\rho + 2\rho^2 + \rho^3)] \tag{4.67}$$

where $2\rho = (c_0 + c_{s,p})r_1$. It is important to note that the *radial* part of the potentials can be also evaluated in spheroidal coordinates by choosing r_1 = fixed as Roothaan (1951a) did. Once the potentials are known, the basic two-electron integrals needed are easily evaluated in spherical coordinates. The results are simple functions of the orbital exponents:

$$(1s^2|1s^2) = \frac{5}{8}c_0 \tag{4.68}$$

$$\begin{cases} (s^2|1s^2) = \dfrac{c_s}{2} - \dfrac{3c_0+c_s}{2}\left(\dfrac{c_s}{c_0+c_s}\right)^5 \\[3mm] (s\,1s|1s^2) = \dfrac{c_0+c_s}{3}S_{s1s}\left[1 - \dfrac{6c_0+c_s}{3c_0+c_s}\left(\dfrac{c_0+c_s}{3c_0+c_s}\right)^3\right] \\[3mm] (s\,1s|s\,1s) = 44\dfrac{c_0^3c_s^5}{(c_0+c_s)^7} \end{cases} \qquad (4.69)$$

$$\begin{cases} (2p_z^2|1s^2) = \dfrac{c_p}{2} - \dfrac{3c_0+c_p}{2}\left(\dfrac{c_p}{c_0+c_p}\right)^5 \\[3mm] (1s\,2p_z|1s2p_z) = \dfrac{28}{3}\dfrac{c_0^3c_p^5}{(c_0+c_p)^7} \end{cases} \qquad (4.70)$$

where the nonorthogonality integral $S_{s1s} = \langle 1s|s\rangle = S$ is given by Equation 3.68.

Then:

$$\begin{cases} J = (2s^2|1s^2) = (1-S^2)^{-1}[(s^2|1s^2) + S^2(1s^2|1s^2) - 2S(s\,1s|1s^2)] \\[2mm] K = (1s\,2s|1s\,2s) = (1-S^2)^{-1}[(s\,1s|s\,1s) + S^2(1s^2|1s^2) - 2S(s\,1s|1s^2)] \end{cases} \qquad (4.71)$$

$$J' = (2p_z^2|1s^2), \quad K' = (1s\,2p_z|1s\,2p_z) \qquad (4.72)$$

As a final example, we give below the calculation of J' and K'.

(i) Evaluation of J'

$$J' = (2p_z^2|J_{1s^2}) = \frac{4}{3}c_p^5\int_0^\infty dr\,\exp(-2c_p r)[r^3 - \exp(-2c_0 r)(r^3 + c_0 r^4)]$$

$$= \frac{4}{3}c_p^5\left\{\frac{3\times2}{(2c_p)^4} - \left[\frac{3\times2}{(2c_0+2c_p)^4} + c_0\frac{4\times3\times2}{(2c_0+2c_p)^5}\right]\right\}$$

$$= \frac{c_p}{2} - \frac{3c_0+c_p}{2}\left(\frac{c_p}{c_0+c_p}\right)^5 \qquad (4.73)$$

(ii) Evaluation of K'

$$K' = (1s\,2p_z|J_{1s2p_z}) = \frac{c_0^{3/2} c_p^{5/2}}{\pi} \int d\mathbf{r}\exp\left[-(c_0+c_p)r\right]$$

$$\times r\cos\theta J_{1s2p_z}(r,\cos\theta)$$

$$= \frac{512}{3}\frac{c_0^3 c_p^5}{(c_0+c_p)^7}\int_0^\infty d\rho \exp(-2\rho)[\rho - \exp(-2\rho)(\rho+2\rho^2+2\rho^3+\rho^4]$$

$$= \frac{512}{3}\frac{c_0^3 c_p^5}{(c_0+c_p)^7}\left(\frac{1}{2^2}-\frac{1}{4^2}-2\frac{2}{4^3}-2\frac{3\times2}{4^4}-\frac{4\times3\times2}{4^5}\right)$$

$$= \frac{1}{6}\frac{c_0^3 c_p^5}{(c_0+c_p)^7}(256-64-64-48-24) = \frac{28}{3}\frac{c_0^3 c_p^5}{(c_0+c_p)^7}$$

$$(4.74)$$

5

Spin

In this chapter, starting from the Zeeman effect, we introduce elements of Pauli's formal theory for one- and two-electron spin, with a generalization to the Dirac formula for N-electron spin.

5.1 THE ZEEMAN EFFECT

The study of atomic spectra under high resolution reveals a new feature of the electron: its spin. Even orbitally nondegenerate levels, like the 1s ground state of the H atom, are split into a doublet under the action of a magnetic field $\mathbf{B}$,[1] the splitting being linear in the field strength (Zeeman effect).

To explain this effect, Pauli (1926) postulated for the electron the existence of *two* spin states, α and β, satisfying the eigenvalue equations

$$\hat{S}_z\alpha = \frac{1}{2}\alpha, \quad \hat{S}_z\beta = -\frac{1}{2}\beta \qquad (5.1)$$

where

$$\hat{\mathbf{S}} = \mathbf{i}\hat{S}_x + \mathbf{j}\hat{S}_y + \mathbf{k}\hat{S}_z \qquad (5.2)$$

---

[1] For a magnetic field of 1 T, the splitting for a ground-state H atom is about 10^5 times smaller than the separation of the first orbital levels.

Methods of Molecular Quantum Mechanics: An Introduction to Electronic Molecular Structure
Valerio Magnasco
© 2009 John Wiley & Sons, Ltd

is the spin vector operator (analogous to the orbital angular momentum vector operator $\hat{\mathbf{L}}$, but whose observable values can be half-integer as well). It is assumed that $\hat{\mathbf{S}}$ and its components satisfy the same commutation rules as those of the angular momentum operators.[2]

The two spin states are assumed to be kets normalized to 1 and orthogonal to each other:

$$\langle \alpha | \alpha \rangle = \langle \beta | \beta \rangle = 1, \quad \langle \alpha | \beta \rangle = \langle \beta | \alpha \rangle = 0 \tag{5.3}$$

It may sometimes be useful to associate with spin the *formal* variable s (as distinct from $\mathbf{r}$, the triplet of coordinates specifying the position in space of the electron) and write Equation 5.3 formally as

$$\int ds\, \alpha^*(s)\alpha(s) = \int ds\, \beta^*(s)\beta(s) = 1, \quad \int ds\, \alpha^*(s)\beta(s) = \int ds\, \beta^*(s)\alpha(s) = 0 \tag{5.4}$$

An electron with spin in a uniform magnetic field $\mathbf{B}$ ($B_x = B_y = 0$, $B_z = B$) acquires a potential energy

$$V(s) = -\hat{\boldsymbol{\mu}}_S \cdot \mathbf{B} = g_e \beta_e B \hat{S}_z \tag{5.5}$$

where $g_e \approx 2$ is the intrinsic electron g-factor (the so-called anomaly of spin), β_e is the Bohr magneton (the unit of magnetic moment), given by

$$\beta_e = \frac{e\hbar}{2mc} = 9.274\,015 \times 10^{-24}\, \mathrm{J\,T^{-1}} \tag{5.6}$$

and $\hat{\boldsymbol{\mu}}_S$, the magnetic moment operator associated with the electron spin $\hat{\mathbf{S}}$, is given by

$$\hat{\boldsymbol{\mu}}_S = -g_e \beta_e \hat{\mathbf{S}} \tag{5.7}$$

The total one-electron Hamiltonian including spin will be

$$\hat{h}(\mathbf{r}, s) = \hat{h}_0(\mathbf{r}) + g_e \beta_e B \hat{S}_z \tag{5.8}$$

---

[2] $[\hat{S}_x, \hat{S}_y] = i\hat{S}_z$, $[\hat{S}_y, \hat{S}_z] = i\hat{S}_x$, $[\hat{S}_z, \hat{S}_x] = i\hat{S}_y$, $[\hat{S}^2, \hat{S}_\kappa] = 0$, $\kappa = x, y, z$.

Let us now introduce a space orbital $\chi_\lambda(\mathbf{r})$ and the two spin-orbitals

$$\begin{cases} \psi_\lambda(\mathbf{x}) = \psi_\lambda(\mathbf{r}, s) = \chi_\lambda(\mathbf{r})\alpha(s) \\ \bar{\psi}_\lambda(\mathbf{x}) = \bar{\psi}_\lambda(\mathbf{r}, s) = \chi_\lambda(\mathbf{r})\beta(s) \end{cases} \tag{5.9}$$

satisfying the eigenvalue equations

$$\begin{cases} \hat{h}_0 \psi_\lambda = \varepsilon_\lambda \psi_\lambda, \quad \hat{h}_0 \bar{\psi}_\lambda = \varepsilon_\lambda \bar{\psi}_\lambda \\ \hat{S}_z \psi_\lambda = \frac{1}{2}\psi, \quad \hat{S}_z \bar{\psi}_\lambda = -\frac{1}{2}\bar{\psi}_\lambda \end{cases} \tag{5.10}$$

The energy level ε_λ hence has a twofold spin degeneracy, which is removed in the presence of the field. Using these two spin-orbitals as an orthonormal basis, the Ritz method shows that the matrix representative of the Hamiltonian $\hat{h}$ over the functions $(\psi_\lambda \, \bar{\psi}_\lambda)$ is already diagonal, the (2×2) secular equation, therefore, having the roots

$$\begin{cases} \varepsilon_1 = \varepsilon_\lambda + \frac{1}{2}g_e\beta_e B \\ \varepsilon_2 = \varepsilon_\lambda - \frac{1}{2}g_e\beta_e B \end{cases} \tag{5.11}$$

which give a Zeeman splitting *linear* in the field B:

$$\Delta\varepsilon = \varepsilon_1 - \varepsilon_2 = g_e\beta_e B \tag{5.12}$$

Transitions from the bottom level ε_2 to the upper one ε_1 are possible by magnetic dipole radiation (Dixon, 1965), so that the electron spin can be reoriented by a photon of energy:

$$h\nu = g_e\beta_e B \tag{5.13}$$

a process originating what is known as electron spin resonance (ESR). A similar process can occur for *nuclei* of spin $\frac{1}{2}$ (^{1}H, ^{13}C), with the order of levels reversed. For ^{1}H:

$$\beta_N = \frac{e\hbar}{2m_p c} = \frac{1}{1836}\frac{e\hbar}{2mc} = \frac{1}{1836}\beta_e \tag{5.14}$$

is the nuclear magneton and $g_N = 5.585$. Connected to the nuclear spin equivalent of (5.13) is the nuclear magnetic resonance (NMR), an experimental technique of great importance nowadays in structural organic chemistry.

5.2 THE PAULI EQUATIONS FOR ONE-ELECTRON SPIN

Until now nothing has been said about the x and y components of the vector operator $\hat{S}$. If we introduce the spin ladder operators

$$\hat{S}_+ = \hat{S}_x + i\hat{S}_y, \quad \hat{S}_- = \hat{S}_x - i\hat{S}_y \tag{5.15}$$

(i being the imaginary unit, $i^2 = -1$),[3] then the analogy with orbital angular momentum suggests for spin the *two-step* ladder of Figure 5.1, where the operators $\hat{S}_+$ and $\hat{S}_-$ respectively step-up (raise) or step-down (lower) the spin functions upon which they act. Therefore, it is intuitive that

$$\begin{cases} \hat{S}_+\alpha = 0 \ (\textit{top} \text{ of the ladder}), \quad \hat{S}_-\alpha = \beta \\ \hat{S}_+\beta = \alpha, \quad \hat{S}_-\beta = 0 \ (\textit{bottom} \text{ of the ladder}) \end{cases} \tag{5.16}$$

By adding and subtracting the corresponding equations, it is easily seen that the two spin states α and β do satisfy the equations

$$\begin{cases} \hat{S}_x\alpha = \frac{1}{2}\beta, \quad \hat{S}_y\alpha = \frac{1}{2}i\beta, \quad \hat{S}_z\alpha = \frac{1}{2}\alpha \\ \hat{S}_x\beta = \frac{1}{2}\alpha, \quad \hat{S}_y\beta = -\frac{1}{2}i\alpha, \quad \hat{S}_z\beta = -\frac{1}{2}\beta \end{cases} \tag{5.17}$$

which are known as Pauli's equations for the one-electron spin and which are of fundamental importance for the whole theory of spin, even in many-electron systems.

Since

$$\hat{S}^2 = \hat{\mathbf{S}} \cdot \hat{\mathbf{S}} = \hat{S}_x^2 + \hat{S}_y^2 + \hat{S}_z^2 \tag{5.18}$$

---

[3] $\hat{S}_+$ and $\hat{S}_-$ have the commutation properties $[\hat{S}_z, \hat{S}_+] = \hat{S}_+$, $[\hat{S}_z, \hat{S}_-] = -\hat{S}_-$, $[\hat{S}^2, \hat{S}_\pm] = 0$.

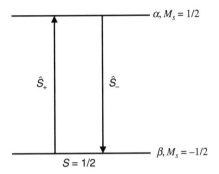

Figure 5.1 The two-step ladder for spin $\frac{1}{2}$

using Equations 5.17 twice it is easily seen that

$$\hat{S}^2\alpha = \frac{3}{4}\alpha, \quad \hat{S}^2\beta = \frac{3}{4}\beta \tag{5.19}$$

so that the two spin states α and β have the *same* eigenvalue with respect to the operator $\hat{S}^2$ ($S(S + 1)$ with $S = \frac{1}{2}$, doublet) and *opposite* eigenvalue ($M_S = \frac{1}{2}$, $M_S = -\frac{1}{2}$) with respect to $\hat{S}_z$. The name doublet (*two* independent spin functions) comes from the so-called spin multiplicity $2S + 1$.

5.3 THE DIRAC FORMULA FOR N-ELECTRON SPIN

For $N = 2$ electrons, we can write the $2^2 = 4$ *products* of two-electron spin functions (which are already eigenfunctions of the operator $\hat{S}_z$):

$$\begin{cases} \quad \alpha\alpha, \quad \alpha\beta, \quad \beta\alpha, \quad \beta\beta \\ M_S = 1 \quad\;\; 0 \quad\;\; 0 \quad -1 \end{cases} \tag{5.20}$$

where electron labels are always assumed in dictionary order, namely:

$$\alpha_1\alpha_2, \; \alpha_1\beta_2, \; \beta_1\alpha_2, \; \beta_1\beta_2 \tag{5.21}$$

and, therefore, are omitted for brevity.

The proper spin *eigenstates* for the $N = 2$ electron spin problem will satisfy the eigenvalue equation

$$\hat{S}^2 \eta = S(S+1)\eta \qquad (5.22)$$

with total spin $S = 0$ (singlet) or $S = 1$ (triplet). To find the spin eigenstates for $N = 2$ it will be sufficient to act upon the set (5.20) with the operator $\hat{S}^2$, finding its matrix representative S^2 over the orthonormal basis set of the four product functions, and diagonalizing it to get its eigenvalues and eigenvectors.

Noting that

$$\hat{S}^2 = \hat{\mathbf{S}} \cdot \hat{\mathbf{S}} = (\hat{\mathbf{S}}_1 + \hat{\mathbf{S}}_2) \cdot (\hat{\mathbf{S}}_1 + \hat{\mathbf{S}}_2) = \hat{S}_1^2 + \hat{S}_2^2 + 2\hat{\mathbf{S}}_1 \cdot \hat{\mathbf{S}}_2 \qquad (5.23)$$

where $\hat{S}_1^2$ and $\hat{S}_2^2$ are one-electron spin operators and $\hat{\mathbf{S}}_1 \cdot \hat{\mathbf{S}}_2$ a two-electron spin operator, we have for the scalar product

$$\begin{aligned}
\hat{\mathbf{S}}_1 \cdot \hat{\mathbf{S}}_2 &= (i\hat{S}_{x1} + j\hat{S}_{y1} + k\hat{S}_{z1}) \cdot (i\hat{S}_{x2} + j\hat{S}_{y2} + k\hat{S}_{z2}) \\
&= \hat{S}_{x1}\hat{S}_{x2} + \hat{S}_{y1}\hat{S}_{y2} + \hat{S}_{z1}\hat{S}_{z2}
\end{aligned} \qquad (5.24)$$

Repeated application of Pauli's equations (5.17) shows, for instance, that

$$(\hat{\mathbf{S}}_1 \cdot \hat{\mathbf{S}}_2)\alpha\beta = \left(\frac{1}{4} + \frac{1}{4}\right)\beta\alpha - \frac{1}{4}\alpha\beta \qquad (5.25)$$

so that

$$\hat{S}^2\alpha\beta = \left(\frac{3}{4} + \frac{3}{4} - \frac{2}{4}\right)\alpha\beta + 2\left(\frac{1}{4} + \frac{1}{4}\right)\beta\alpha = \alpha\beta + \beta\alpha \qquad (5.26)$$

Completing the process for all product functions (5.20), we obtain for matrix S^2 the block-diagonal form:

$$S^2 = \begin{pmatrix} 2 & 0 & 0 & 0 \\ 0 & 1 & 1 & 0 \\ 0 & 1 & 1 & 0 \\ 0 & 0 & 0 & 2 \end{pmatrix} \qquad (5.27)$$

The (2×2) inner block gives the secular equation:

$$\begin{vmatrix} 1-\lambda & 1 \\ 1 & 1-\lambda \end{vmatrix} = 0 \quad \lambda = S(S+1) \tag{5.28}$$

having the roots

$$\begin{cases} \lambda_4 = 0 \Rightarrow S = 0 \\ \lambda_2 = 2 \Rightarrow S = 1 \end{cases} \tag{5.29}$$

Solving the associated linear homogeneous system gives

$$\begin{cases} \eta_4 = \dfrac{1}{\sqrt{2}}(\alpha\beta - \beta\alpha) \quad (S = 0) \\[2ex] \eta_2 = \dfrac{1}{\sqrt{2}}(\alpha\beta + \beta\alpha) \quad (S = 1) \end{cases} \tag{5.30}$$

for the two eigenvectors. Therefore, we obtain the $N = 2$ spin eigenstates (N_α is the number of electrons with spin α and N_β that of electrons with spin β, with $N_\alpha + N_\beta = N$) collected in Table 5.1.

In conclusion, we see that the eigenstates of the two-electron spin problem have definite symmetry properties under the interchange $1 \leftrightarrow 2$ (or $\alpha \leftrightarrow \beta$), the triplet being *symmetric* and the singlet *antisymmetric* under electron (spin) interchange.

From Equation 5.26 we further note that we can write for $\hat{S}^2$

$$\hat{S}^2 = \hat{I} + \hat{P}_{12} = \hat{I} + \hat{P}_{\alpha\beta} \tag{5.31}$$

Table 5.1 Eigenstates of the two-electron problem

Eigenstates η	S	$M_S = (N_\alpha - N_\beta)/2$
$\eta_1 = \alpha\alpha$	1	1
$\eta_2 = \dfrac{1}{\sqrt{2}}(\alpha\beta + \beta\alpha)$	1	0
$\eta_3 = \beta\beta$	1	−1
$\eta_4 = \dfrac{1}{\sqrt{2}}(\alpha\beta - \beta\alpha)$	0	0

a formula due to Dirac (1929), which can be easily generalized to the
N-electron spin problem as (Löwdin, 1955)

$$\hat{S}^2 = \frac{N}{4}(4-N)\hat{I} + \sum_{\kappa < \lambda} \hat{P}_{\kappa\lambda} \qquad (5.32)$$

where $\hat{I}$ is the identity operator and $\hat{P}_{\kappa\lambda}$ interchanges spin state κ with spin
state λ. Dirac's formula (5.32) allows one to avoid repeated use of Pauli's
equations (5.17) and to calculate in a very simple way the effect of $\hat{S}^2$ upon
many-electron (spin) functions. In the case of Slater determinants, which
we shall meet later on, the operator (5.32) will act only upon the spin part
of the spin-orbital functions, leaving the space orbitals unchanged.

The eigenstates of $\hat{S}^2$ for the N-electron spin problem are states of
definite total spin S (≥ 0) satisfying simultaneously the two eigenvalue
equations

$$\hat{S}^2\eta = S(S+1)\eta, \quad \hat{S}_z\eta = M_S\eta \qquad (5.33)$$

where

$$\begin{cases} S = 0, 1, 2, \ldots, \dfrac{N}{2} & \text{for } N = \text{even} \\[2mm] S = \dfrac{1}{2}, \dfrac{3}{2}, \dfrac{5}{2}, \ldots, \dfrac{N}{2} & \text{for } N = \text{odd} \end{cases} \qquad (5.34)$$

$$M_S = \frac{N_\alpha - N_\beta}{2} \qquad (5.35)$$

In terms of the spin multiplicity $2S+1$, we have

$$N = \text{even} \Rightarrow \text{singlets, triplets, quintets, } \ldots \qquad (5.36)$$

$$N = \text{odd} \Rightarrow \text{doublets, quartets, sextets, } \ldots \qquad (5.37)$$

The number f_S^N of linearly independent spin eigenstates of given
S for the N-electron spin system is given by a formula due to Wigner
(1959):

$$f_S^N = \frac{(2S+1)N!}{\left(\dfrac{N}{2}-S\right)!\left(\dfrac{N}{2}+S+1\right)!} \qquad (5.38)$$

As an example, for the three-electron system, Wigner's formula gives

$$
\begin{cases}
N = 3, \ S = \dfrac{1}{2} \Rightarrow f^3_{1/2} = 2 \\[2mm]
N = 3, \ S = \dfrac{3}{2} \Rightarrow f^3_{3/2} = 1
\end{cases}
\tag{5.39}
$$

so that we have *two* distinct doublets (four functions) and *one* quartet (four functions as well).

6

Antisymmetry of Many-electron Wavefunctions

The effects of symmetry on the many-electron wavefunctions arise from the physical identity and, therefore, from the indistinguishability of the electrons, and are of fundamental importance in the treatment of the many-particle systems. Use of suitable density functions allows us to pass from the abstract $4N$-dimensional mathematical space of the N-electron wavefunctions to the three-dimensional + spin space where physical experiments are done (McWeeny, 1960).

6.1 ANTISYMMETRY REQUIREMENT AND THE PAULI PRINCIPLE

Let x_1 and x_2 be two *fixed* points in the space–spin space and $\Psi(x_1, x_2)$ a normalized two-electron wavefunction. Then, since electrons are *identical* particles, we have from first principles that

$$
\begin{cases}
|\Psi(x_1, x_2)|^2 \, dx_1 \, dx_2 = \text{probability of finding electron 1} \\
\quad \text{at } dx_1 \text{ and electron 2 at } dx_2 \\
= |\Psi(x_2, x_1)|^2 \, dx_1 \, dx_2 = \text{probability of finding electron 2} \\
\quad \text{at } dx_1 \text{ and electron 1 at } dx_2
\end{cases}
\tag{6.1}
$$

Methods of Molecular Quantum Mechanics: An Introduction to Electronic Molecular Structure
Valerio Magnasco
© 2009 John Wiley & Sons, Ltd

Therefore, it follows that

$$|\Psi(\mathbf{x}_2, \mathbf{x}_1)|^2 = |\Psi(\mathbf{x}_1, \mathbf{x}_2)|^2 \Rightarrow \Psi(\mathbf{x}_2, \mathbf{x}_1) = \pm\Psi(\mathbf{x}_1, \mathbf{x}_2) \qquad (6.2)$$

and the wavefunction must be symmetric (+ sign) or antisymmetric (− sign) in the interchange of the space–spin coordinates of the two electrons.

The Pauli principle states that 'electrons must be described only by antisymmetric wavefunctions', namely:

$$\Psi(\mathbf{x}_2, \mathbf{x}_1) = -\Psi(\mathbf{x}_1, \mathbf{x}_2) \qquad (6.3)$$

which is *Pauli's antisymmetry principle* in the form given by Dirac (1929). This formulation includes the well-known *exclusion principle* for electrons in the *same* orbital with the *same* spin:

$$\Psi(\mathbf{x}_1, \mathbf{x}_2) = \psi_\lambda(\mathbf{x}_1)\psi_\lambda(\mathbf{x}_2) - \psi_\lambda(\mathbf{x}_1)\psi_\lambda(\mathbf{x}_2) = 0 \qquad (6.4)$$

where, as usual, we take electrons in dictionary order and interchange spin-orbitals only. It is, instead, allowed to put two electrons in the *same* orbital with *different* spin:

$$\begin{aligned}
\Psi(\mathbf{x}_1, \mathbf{x}_2) &= \psi_\lambda(\mathbf{x}_1)\bar{\psi}_\lambda(\mathbf{x}_2) - \bar{\psi}_\lambda(\mathbf{x}_1)\psi_\lambda(\mathbf{x}_2) \\
&= \begin{vmatrix} \psi_\lambda(\mathbf{x}_1) & \bar{\psi}_\lambda(\mathbf{x}_1) \\ \psi_\lambda(\mathbf{x}_2) & \bar{\psi}_\lambda(\mathbf{x}_2) \end{vmatrix} = |\psi_\lambda(\mathbf{x}_1)\bar{\psi}_\lambda(\mathbf{x}_2)|
\end{aligned} \qquad (6.5)$$

where the Pauli-allowed wavefunction can be written in the form of a determinant of order 2 with electrons as rows and spin-orbitals as columns.[1] This determinant is known as *Slater determinant*, since Slater (1929) was the first to introduce this notation in his quantum mechanical treatment of atomic multiplets without using group theory.

If the spin-orbitals are assumed to be orthonormal, then we shall write the *normalized* Slater determinant within a vertical double bar as

$$\Psi(\mathbf{x}_1, \mathbf{x}_2) = \frac{1}{\sqrt{2}}|\psi_\lambda(\mathbf{x}_1)\bar{\psi}_\lambda(\mathbf{x}_2)| = ||\psi_\lambda(\mathbf{x}_1)\bar{\psi}_\lambda(\mathbf{x}_2)|| \qquad (6.6)$$

---

[1] Recall that we use ψ_λ to denote a spin-orbital with spin α and $\bar{\psi}_\lambda$ a spin-orbital with spin β.

6.2   SLATER DETERMINANTS

Generalizing to the N-electron system, we shall write the N-electron wavefunction Ψ satisfying Pauli's principle as

$$\Psi(x_1, x_2, \ldots, x_N) = \frac{1}{\sqrt{N!}} \begin{vmatrix} \psi_1(x_1) & \psi_2(x_1) & \cdots & \psi_N(x_1) \\ \psi_1(x_2) & \psi_2(x_2) & \cdots & \psi_N(x_2) \\ \cdots & \cdots & \cdots & \cdots \\ \psi_1(x_N) & \psi_2(x_N) & \cdots & \psi_N(x_N) \end{vmatrix}$$

$$= \|\psi_1 \psi_2 \cdots \psi_N\| \tag{6.7}$$

a Slater determinant of order N, where *rows* denote space–spin coordinates of the electrons ($x = rs$) and *columns* the spin-orbital functions. If the latter are orthonormal, then Ψ of Equation 6.7 is normalized to 1:

$$\langle \Psi | \Psi \rangle = \int dx_1\, dx_2 \ldots dx_N \Psi^*(x_1, x_2, \ldots, x_N)\Psi(x_1, x_2, \ldots, x_N) = 1 \tag{6.8}$$

The elementary properties of determinants introduced in Chapter 2 show that (6.7) can be equally well written as

$$\Psi(x_1, x_2, \ldots, x_N) = \frac{1}{\sqrt{N!}} \begin{vmatrix} \psi_1(x_1) & \psi_1(x_2) & \cdots & \psi_1(x_N) \\ \psi_2(x_1) & \psi_2(x_2) & \cdots & \psi_2(x_N) \\ \cdots & \cdots & \cdots & \cdots \\ \psi_N(x_1) & \psi_N(x_2) & \cdots & \psi_N(x_N) \end{vmatrix} \tag{6.9}$$

where we now choose spin-orbitals as rows and electrons as columns, having interchanged rows with columns in the original definition (6.7).

Furthermore, it is easily seen that Ψ does satisfy Pauli's antisymmetry principle:

$$\Psi(x_2, x_1, \ldots, x_N) = \frac{1}{\sqrt{N!}} \begin{vmatrix} \psi_1(x_2) & \psi_2(x_2) & \cdots & \psi_N(x_2) \\ \psi_1(x_1) & \psi_2(x_1) & \cdots & \psi_N(x_1) \\ \cdots & \cdots & \cdots & \cdots \\ \psi_1(x_N) & \psi_2(x_N) & \cdots & \psi_N(x_N) \end{vmatrix}$$

$$= -\Psi(x_1, x_2, \ldots, x_N) \tag{6.10}$$

since this is equivalent to interchange of two rows in the determinant (6.7) and the determinant changes sign.

Also, Pauli's exclusion principle is satisfied by (6.7):

$$\Psi(x_1, x_2, \ldots, x_N) = \frac{1}{\sqrt{N!}} \begin{vmatrix} \psi_1(x_1) & \psi_1(x_1) & \cdots & \psi_N(x_1) \\ \psi_1(x_2) & \psi_1(x_2) & \cdots & \psi_N(x_2) \\ \cdots & \cdots & \cdots & \cdots \\ \psi_1(x_N) & \psi_1(x_N) & \cdots & \psi_N(x_N) \end{vmatrix} = 0 \quad (6.11)$$

since (6.11) now has two identical columns and the determinant vanishes.

For the N-electron system, the wavefunction Ψ is antisymmetric if it is left invariant by an *even* number of electron interchanges (permutations) and it changes sign with an *odd* number of interchanges. The probability density (as the Hamiltonian operator) is instead left *unchanged* after any number of interchanges among the electrons.[2]

A single Slater determinant is often sufficient as a first approximation for closed-shell systems ($S = M_S = 0$), but some components of the open-shell systems, like the $M_S = 0$ components of the triplet states $1s2s(^3S)$ of He or $1\sigma_g1\sigma_u(^3\Sigma_u^+)$ of H_2, require a linear combination of Slater determinants to get a state of definite S.

A few atomic and molecular examples are given below.

(i) Ground state of the He atom ($N = 2$)

$$\Psi(1s^2, {}^1S) = ||1s\,1\bar{s}|| \quad S = M_S = 0$$

(ii) Ground state of the Be atom ($N=4$)

$$\Psi(1s^22s^2, {}^1S) = ||1s\,1\bar{s}\,2s\,2\bar{s}|| \quad S = M_S = 0$$

(iii) Ground state of the H_2 molecule ($N=2$)

$$\Psi(1\sigma_g^2, {}^1\Sigma_g^+) = ||1\sigma_g1\bar{\sigma}_g|| \quad S = M_S = 0$$

where

$$1\sigma_g = \frac{a+b}{\sqrt{2+2S}} \quad a = 1s_A, \; b = 1s_B \quad (6.12)$$

is the bonding molecular orbital doubly occupied by the electrons.

---

[2] Recall that we simply interchange spin-orbitals between the electrons kept in dictionary order.

(iv) Ground state of the LiH molecule ($N = 4$)

$$\Psi(1\sigma^2\, 2\sigma^2,\, ^1\Sigma^+) = ||1\sigma\, 1\bar{\sigma}\, 2\sigma\, 2\bar{\sigma}||\ S = M_S = 0$$

where 1σ and 2σ are the first two MOs obtained, in a first approximation, by the linear combination of the basic AOs (1s, 2s) on Li and (1s) on H.

All these electronic states are described by a single Slater determinant. But:

(v) First excited 1s2s state of the He atom ($N = 2$)

$$\Psi(1s\, 2s,\, ^3S) = \begin{cases} ||1s\, 2s|| & S = 1,\quad M_S = 1 \\[2mm] \dfrac{1}{\sqrt{2}}\left[||1s2\bar{s}|| + ||1\bar{s}\, 2s||\right] & S = 1,\quad M_S = 0 \\[2mm] ||1\bar{s}\, 2\bar{s}|| & S = 1,\quad M_S = -1 \end{cases}$$

Only the components of the triplet with $|M_S| = S$ are described by a single Slater determinant, whereas the component with $M_S = 0$ requires *two* Slater determinants.

6.3 DISTRIBUTION FUNCTIONS

The passage from the abstract mathematical space of the N-electron wavefunction Ψ to the physical four-dimensional space including spin is done by the distribution functions. Since the Hamiltonian operator for a many-electron atom or molecule contains only symmetrical sums of one-electron and two-electron operators, the most important distribution functions for us will be those involving *one* and *two* electrons respectively.

6.3.1 One- and Two-electron Distribution Functions

The distribution functions are the diagonal elements of the one- and two-electron *density matrices* (McWeeny, 1960) defined as

$$\rho_1(\mathbf{x}_1; \mathbf{x}'_1) = N \int d\mathbf{x}_2\, d\mathbf{x}_3 \ldots d\mathbf{x}_N\, \Psi(\mathbf{x}_1, \mathbf{x}_2, \ldots, \mathbf{x}_N)\Psi^*(\mathbf{x}'_1, \mathbf{x}_2, \ldots, \mathbf{x}_N)$$

$$(6.13)$$

$$\rho_2(\mathbf{x}_1, \mathbf{x}_2; \mathbf{x'}_1, \mathbf{x'}_2) = N(N-1) \int d\mathbf{x}_3 \ldots d\mathbf{x}_N \, \Psi(\mathbf{x}_1, \mathbf{x}_2, \mathbf{x}_3, \ldots, \mathbf{x}_N)$$
$$\times \Psi^*(\mathbf{x'}_1, \mathbf{x'}_2, \mathbf{x}_3, \ldots, \mathbf{x}_N) \tag{6.14}$$

where the first set of variables in ρ_1 and ρ_2 comes from Ψ and the second from Ψ^*. Functions (6.13) and (6.14) have only a mathematical meaning, and are needed when a differential operator (like ∇^2) acts on the ρs.[3] All physical meaning is carried instead by their *diagonal elements* $(\mathbf{x'}_1 = \mathbf{x}_1)$ and $(\mathbf{x'}_1 = \mathbf{x}_1, \mathbf{x'}_2 = \mathbf{x}_2)$, which are the one- and two-electron *distribution functions*

$$\rho_1(\mathbf{x}_1; \mathbf{x}_1) = N \int d\mathbf{x}_2 \, d\mathbf{x}_3 \ldots d\mathbf{x}_N \, \Psi(\mathbf{x}_1, \mathbf{x}_2, \ldots, \mathbf{x}_N)\Psi^*(\mathbf{x}_1, \mathbf{x}_2, \ldots, \mathbf{x}_N)$$
$$\tag{6.15}$$

$$\rho_2(\mathbf{x}_1, \mathbf{x}_2; \mathbf{x}_1, \mathbf{x}_2) = N(N-1) \int d\mathbf{x}_3 \ldots d\mathbf{x}_N \, \Psi(\mathbf{x}_1, \mathbf{x}_2, \mathbf{x}_3, \ldots, \mathbf{x}_N)$$
$$\times \Psi^*(\mathbf{x}_1, \mathbf{x}_2, \mathbf{x}_3, \ldots, \mathbf{x}_N) \tag{6.16}$$

having the following conservation properties:

$$\int d\mathbf{x}_1 \, \rho_1(\mathbf{x}_1; \mathbf{x}_1) = N \tag{6.17}$$

the total number of electrons and

$$\int d\mathbf{x}_2 \, \rho_2(\mathbf{x}_1, \mathbf{x}_2; \mathbf{x}_1, \mathbf{x}_2) = (N-1)\rho_1(\mathbf{x}_1; \mathbf{x}_1),$$
$$\int d\mathbf{x}_1 \, d\mathbf{x}_2 \, \rho_2(\mathbf{x}_1, \mathbf{x}_2; \mathbf{x}_1, \mathbf{x}_2) = N(N-1) \tag{6.18}$$

the total number of indistinct pairs.

The physical meaning of the distribution functions (6.15) and (6.16) is as follows:

$$\rho_1(\mathbf{x}_1; \mathbf{x}_1) \, d\mathbf{x}_1 = \text{probability of finding } an \text{ electron at } d\mathbf{x}_1 \tag{6.19}$$

---

[3] We recall that the operator ∇^2 acts only on Ψ and not on Ψ^*.

$\rho_2(\mathbf{x}_1, \mathbf{x}_2; \mathbf{x}_1, \mathbf{x}_2)\, d\mathbf{x}_1\, d\mathbf{x}_2$ = probability of finding *an* electron at $d\mathbf{x}_1$ and, simultaneously, *another* electron at $d\mathbf{x}_2$

$$(6.20)$$

In this way, the distribution functions take into account from the beginning the fact that electrons are physically identical particles. The two-electron (or *pair*) distribution function is very important in the study of the correlation between electrons (McWeeny, 1960), but this point will not be analysed further here. Instead, we shall further examine in more detail the properties of the one-electron distribution function.

6.3.2 Electron and Spin Densities

In its most general form, the one-electron distribution (or density) function (6.15) can be written by separating its space from spin parts as follows:

$$\rho_1(\mathbf{r}_1 s_1; \mathbf{r}_1 s_1) = \rho_1^{\alpha\alpha}(\mathbf{r}_1; \mathbf{r}_1)\alpha(s_1)\alpha^*(s_1) + \rho_1^{\beta\beta}(\mathbf{r}_1; \mathbf{r}_1)\beta(s_1)\beta^*(s_1)$$

$$+ \rho_1^{\alpha\beta}(\mathbf{r}_1; \mathbf{r}_1)\alpha(s_1)\beta^*(s_1) + \rho_1^{\beta\alpha}(\mathbf{r}_1; \mathbf{r}_1)\beta(s_1)\alpha^*(s_1)$$

$$(6.21)$$

where the ρs are now space functions only. By integrating over spin, the last two terms in (6.21) vanish, because of the orthogonality of the spin functions, and we are left with the *spinless* quantities

$$\rho_1^{\alpha\alpha}(\mathbf{r}_1; \mathbf{r}_1) = \rho_1^{\alpha}(\mathbf{r}_1; \mathbf{r}_1), \quad \rho_1^{\beta\beta}(\mathbf{r}_1; \mathbf{r}_1) = \rho_1^{\beta}(\mathbf{r}_1; \mathbf{r}_1) \qquad (6.22)$$

having the following evident physical meaning:

$\rho_1^{\alpha}(\mathbf{r}_1; \mathbf{r}_1)\, d\mathbf{r}_1$ = probability of finding at $d\mathbf{r}_1$ an electron with spin α

$$(6.23)$$

$\rho_1^{\beta}(\mathbf{r}_1; \mathbf{r}_1)\, d\mathbf{r}_1$ = probability of finding at $d\mathbf{r}_1$ an electron with spin β

$$(6.24)$$

with

$$\int d\mathbf{r}_1\, \rho_1^{\alpha}(\mathbf{r}_1; \mathbf{r}_1) = N_\alpha \qquad (6.25)$$

the number of electrons with spin α and

$$\int d\mathbf{r}_1 \, \rho_1^\beta(\mathbf{r}_1; \mathbf{r}_1) = N_\beta \tag{6.26}$$

the number of electrons with spin β. Therefore, the sum

$$P(\mathbf{r}_1; \mathbf{r}_1) = \rho_1^\alpha(\mathbf{r}_1; \mathbf{r}_1) + \rho_1^\beta(\mathbf{r}_1; \mathbf{r}_1) \tag{6.27}$$

is the *electron density*, as observed, for example, from the X-ray diffraction spectra of polycyclic hydrocarbons (Bacon, 1969), with

$P(\mathbf{r}_1; \mathbf{r}_1) \, d\mathbf{r}_1 =$ probability of finding at $d\mathbf{r}_1$ an electron with *either* spin

$$\tag{6.28}$$

whereas the difference

$$Q(\mathbf{r}_1; \mathbf{r}_1) = \rho_1^\alpha(\mathbf{r}_1; \mathbf{r}_1) - \rho_1^\beta(\mathbf{r}_1; \mathbf{r}_1) \tag{6.29}$$

is the *spin density*, with

$Q(\mathbf{r}_1; \mathbf{r}_1) \, d\mathbf{r}_1 =$ probability of finding at $d\mathbf{r}_1$ an *excess* of spin α over spin β

$$\tag{6.30}$$

P and Q satisfy the following conservation relations:

$$\int d\mathbf{r}_1 \, P(\mathbf{r}_1; \mathbf{r}_1) = N_\alpha + N_\beta = N \tag{6.31}$$

the total number of electrons and

$$\int d\mathbf{r}_1 \, Q(\mathbf{r}_1; \mathbf{r}_1) = N_\alpha - N_\beta = 2M_S \tag{6.32}$$

where

$$M_S = \frac{N_\alpha - N_\beta}{2} \tag{6.33}$$

is the eigenvalue of the $\hat{S}_z$ operator

$$\hat{S}_z \Psi = M_S \Psi \qquad (6.34)$$

As a simple example, let us consider now the two-electron wavefunction Ψ describing the bond between atoms A and B arising from the double occupancy of the (normalized) *bonding* MO $\phi(\mathbf{r})$:

$$\Psi(\mathbf{x}_1, \mathbf{x}_2) = \|\phi\bar{\phi}\| = \phi(\mathbf{r}_1)\phi(\mathbf{r}_2)\frac{1}{\sqrt{2}}[\alpha(s_1)\beta(s_2) - \beta(s_1)\alpha(s_2)] \qquad (6.35)$$

with

$$\phi(\mathbf{r}) = \chi_A(\mathbf{r})c_A + \chi_B(\mathbf{r})c_B = \frac{\chi_A(\mathbf{r}) + \lambda\chi_B(\mathbf{r})}{\sqrt{1 + \lambda^2 + 2\lambda S}} \qquad (6.36)$$

where χ_A and χ_B are two basis AOs centred on A and B respectively, normalized but not orthogonal to each other, and $\lambda = c_B/c_A$ is the *polarity* parameter of the MO. The integral

$$S = \langle \chi_A | \chi_B \rangle = \int d\mathbf{r}\, \chi_A^*(\mathbf{r})\chi_B(\mathbf{r}) \qquad (6.37)$$

describes the *overlap* between the AOs χ_A and χ_B and, therefore, is called the overlap integral.[4]

Using definition (6.15) omitting for brevity the suffix 1, useless at the moment, the integration over $d\mathbf{x}_2$ gives the one-electron distribution function

$$\rho_1(\mathbf{x}; \mathbf{x}) = \phi(\mathbf{r})\phi^*(\mathbf{r})[\alpha(s)\alpha^*(s) + \beta(s)\beta^*(s)] \qquad (6.38)$$

so that

$$\rho_1^\alpha(\mathbf{r}; \mathbf{r}) = \rho_1^\beta(\mathbf{r}; \mathbf{r}) = \phi(\mathbf{r})\phi^*(\mathbf{r}) = R(\mathbf{r}; \mathbf{r}) \qquad (6.39)$$

---

[4] S depends on the internuclear distance R, being one at $R = 0$ and zero at $R = \infty$.

The coefficients of $\alpha\alpha^*$ and $\beta\beta^*$ are equal for the doubly occupied MO and are usually denoted by R in MO theory. We then obtain for the electron and spin densities of our bond orbital

$$P(\mathbf{r};\mathbf{r}) = \rho_1^\alpha(\mathbf{r};\mathbf{r}) + \rho_1^\beta(\mathbf{r};\mathbf{r}) = 2R(\mathbf{r};\mathbf{r}) = 2\phi(\mathbf{r})\phi^*(\mathbf{r}) \qquad (6.40)$$

$$Q(\mathbf{r};\mathbf{r}) = \rho_1^\alpha(\mathbf{r};\mathbf{r}) - \rho_1^\beta(\mathbf{r};\mathbf{r}) = 0 \qquad (6.41)$$

as expected for an MO doubly occupied by electrons with opposite spin and $M_S = 0$.

The electron density (6.40) can be further analysed in terms of elementary contributions from the AOs, giving the so-called *population analysis*, which shows how the electrons are distributed between the different atomic orbitals in the molecule. We obtain

$$P(\mathbf{r};\mathbf{r}) = q_A\chi_A^2(\mathbf{r}) + q_B\chi_B^2(\mathbf{r}) + q_{AB}\frac{\chi_A(\mathbf{r})\chi_B(\mathbf{r})}{S} + q_{BA}\frac{\chi_B(\mathbf{r})\chi_A(\mathbf{r})}{S} \qquad (6.42)$$

from (6.36), where $\chi_A^2(\mathbf{r})$ and $\chi_B^2(\mathbf{r})$ are the *atomic densities* and $(\chi_A(\mathbf{r})\chi_B(\mathbf{r}))/S$ and $(\chi_B(\mathbf{r})\chi_A(\mathbf{r}))/S$ are the *overlap densities*, all normalized to 1, while the coefficients

$$q_A = \frac{2}{1+\lambda^2+2\lambda S}, \quad q_B = \frac{2\lambda^2}{1+\lambda^2+2\lambda S} \qquad (6.43)$$

are the atomic charges and

$$q_{AB} = q_{BA} = \frac{2\lambda S}{1+\lambda^2+2\lambda S} \qquad (6.44)$$

are the *overlap charges*. The charges are normalized so that

$$q_A + q_B + q_{AB} + q_{BA} = \frac{2+2\lambda^2+4\lambda S}{1+\lambda^2+2\lambda S} = 2 \qquad (6.45)$$

the total number of electrons in the bond orbital $\phi(\mathbf{r})$.

For a *homopolar* bond, $\lambda = 1$:

$$q_A = q_B = \frac{1}{1+S}, \quad q_{AB} = q_{BA} = \frac{S}{1+S} \qquad (6.46)$$

so that for $S > 0$, in the molecule, the charge on atoms is decreased, electrons being transferred in the intermediate region between nuclei to an extent described by q_{AB} and q_{BA}. This reduces internuclear repulsion and means *bonding*.

For a *heteropolar* bond, $\lambda \neq 1$, and we define *gross charges* on A and B as

$$Q_A = q_A + q_{AB} = \frac{2 + 2\lambda S}{1 + \lambda^2 + 2\lambda S} \qquad (6.47)$$

$$Q_B = q_B + q_{BA} = \frac{2\lambda^2 + 2\lambda S}{1 + \lambda^2 + 2\lambda S} \qquad (6.48)$$

and *formal charges* on A and B as

$$\delta_A = 1 - Q_A = \frac{\lambda^2 - 1}{1 + \lambda^2 + 2\lambda S} \qquad (6.49)$$

$$\delta_B = 1 - Q_B = -\frac{\lambda^2 - 1}{1 + \lambda^2 + 2\lambda S} \qquad (6.50)$$

If $\lambda > 1$, then $\delta_A = \delta > 0$ and $\delta_B = -\delta_A = -\delta < 0$ and we have the dipole $A^{+\delta} B^{-\delta}$ (e.g. the LiH molecule). Further examples for the $^1\Sigma_g^+$ ground state and the $^3\Sigma_u^+$ excited triplet state of the H_2 molecule can be found elsewhere (Magnasco, 2007). It is seen there that, in a repulsive state such as triplet H_2 (or ground-state He_2), electrons escape from the interatomic region determining repulsion (*antibonding*) between the interacting atoms, the so-called Pauli repulsion.

Generalization of this analysis to the N-electron determinant is known as Mulliken population analysis (Mulliken, 1955).

6.4   AVERAGE VALUES OF OPERATORS

With the aid of the density functions we can easily evaluate the *average values* of any symmetrical sum of one- and two-electron operators:

$$\left\langle \Psi \left| \sum_{i=1}^{N} \hat{O}_i \right| \Psi \right\rangle = \int d\mathbf{x}_1 \, \hat{O}_1 \rho_1(\mathbf{x}_1; \mathbf{x}_1')|_{\mathbf{x}_1' = \mathbf{x}_1} \qquad (6.51)$$

$$\left\langle \Psi \middle| \sum_{i,j=1(j\neq i)}^{N} \hat{O}_{ij} \middle| \Psi \right\rangle = \int dx_1 \, dx_2 \, \hat{O}_{12} \rho_2(x_1 x_2; x_1 x_2) \qquad (6.52)$$

where Ψ is the N-electron wavefunction normalized to 1 and $\hat{O}_i$ and $\hat{O}_{ij}$ are the one- and two-electron operators respectively. In (6.51) we have used the one-electron density matrix (6.13) instead of the one-electron distribution function, leaving the possibility that $\hat{O}_1$ could be a differential operator (like ∇^2), which acts on the first set of variables in ρ_1 but *not* on the second. The notation in (6.51) should by now be clear: first, let the operator $\hat{O}_1$ act on ρ_1, then put $x_1' = x_1$ in the resulting integrand and integrate over x_1. In (6.52) we have assumed that the operator $\hat{O}_{12}$ is a simple multiplier (like the electron repulsion $1/r_{12}$).

The electronic Hamiltonian $\hat{H}_e$ contains two such symmetrical sums, and its average value over the N-electron wavefunction Ψ can be written as

$$\langle \Psi | \hat{H}_e | \Psi \rangle = \left\langle \Psi \middle| \sum_{i=1}^{N} \hat{h}_i + \frac{1}{2} \sum_{i,j=1(j\neq i)}^{N} \frac{1}{r_{ij}} \middle| \Psi \right\rangle$$

$$= \int dx_1 \, \hat{h}_1 \rho_1(x_1; x_1')|_{x_1'=x_1} + \frac{1}{2} \int dx_1 \, dx_2 \, \frac{1}{r_{12}} \rho_2(x_1 x_2; x_1 x_2) \qquad (6.53)$$

where

$$\hat{h}_1 = -\frac{1}{2}\nabla_1^2 + V_1, \quad V_1 = -\sum_{\alpha} \frac{Z_\alpha}{r_{\alpha 1}} \qquad (6.54)$$

is the one-electron bare nuclei Hamiltonian, V_1 is the attraction of the electron by all nuclei of charge $+Z_\alpha$ in the molecule, and $1/r_{12}$ the electron repulsion.

Hence, the electronic energy E_e consists of the following three terms:

$$E_e = \int dx_1 \left[-\frac{1}{2}\nabla_1^2 \rho_1(x_1; x_1')|_{x_1'=x_1} \right] + \int dx_1 \, V_1 \, \rho_1(x_1; x_1)$$

$$+ \frac{1}{2}\int dx_1 \, dx_2 \, \frac{1}{r_{12}} \rho_2(x_1 x_2; x_1 x_2) \qquad (6.55)$$

which have the following simple, physically transparent, interpretation. The first term in (6.55) is the average kinetic energy of the electron distribution ρ_1, the second is the average potential energy of the electron distribution ρ_1 in the field provided by all nuclei in the molecule, and the third is the average electronic repulsion of an electron pair described by the pair function ρ_2.

Lastly, it is important to stress that the integral in (6.51) would involve $N(N!)^2$ $4N$-dimensional integrations, which are reduced to a simple four-dimensional space–spin integration by use of the density matrix! General rules for the evaluation of the integrals in (6.55) in the general case of two *different* Slater determinants Ψ and Ψ' built from orthonormal spin-orbitals were first given by Slater (1931) and recast in density matrix form by McWeeny (1960).

7

Self-consistent-field Calculations and Model Hamiltonians

Hartree–Fock (HF) theory was developed in the early 1930s by the theoretical physicists Hartree (1928a, 1928b) and Fock (1930) to deal with the quantum mechanical problem of many-particle atomic and molecular systems. Initially developed by Hartree for complex atoms in a numerical form using as Ψ the simple *product* of N spin-orbitals, the theory was put on a more sound basis by Fock, who first solved the variational problem of optimizing the orbitals in a *determinantal* Ψ obeying Pauli's antisymmetry principle. HF theory amounts essentially to finding the best form for the orbitals inside the so-called *independent particle model* (IPM), where an electron in the molecule is assumed to move in the average field provided by all nuclei and the sea of all other electrons. Besides the Coulomb potential J due to all electrons, the antisymmetry requirement imposed by Fock upon the wavefunction Ψ originates a nonlocal exchange potential K.

Since no correlation between the electrons is provided by this model, the HF energy is used to define exactly the *correlation energy* of the system as

$$\text{Correlation energy} = \text{Exact energy of the nonrelativistic Hamiltonian} - \text{HF energy}$$

$$(7.1)$$

Methods of Molecular Quantum Mechanics: An Introduction to Electronic Molecular Structure
Valerio Magnasco
© 2009 John Wiley & Sons, Ltd

The correlation energy amounts to about 1 eV per electron pair and is quite difficult to account for, as we shall briefly outline in Chapter 8. We shall introduce, first, the essential lines of the HF theory for closed shells, with its practical implementation by Hall (1951) and Roothaan (1951b) yielding the so-called self-consistent-field (SCF) method inside the Ritz linear-combination-of-atomic-orbitals (LCAO) approach, and, next, the LCAO method devised by Hückel (1931) to deal in a topological way with the π electrons of conjugated and aromatic hydrocarbons. Hückel theory has recently been used by the author to introduce an elementary model of the chemical bond (Magnasco, 2002, 2003, 2004a, 2005).

7.1 ELEMENTS OF HARTREE–FOCK THEORY FOR CLOSED SHELLS

Let

$$\Psi = \|\psi_1 \psi_2 \ldots \psi_N\| \quad \langle \psi_i | \psi_j \rangle = \delta_{ij} \quad S = M_S = 0 \qquad (7.2)$$

be a normalized single determinant wavefunction of doubly occupied orthonormal spin-orbitals $\{\psi_i\}$. We shall introduce consistently the following notation:

$N =$ number of electrons = number of spin-orbitals $\{\psi_i(\mathbf{x})\}$, $\mathbf{x} = \mathbf{r}s$, $i = 1, 2, \ldots, N$

$n = N/2$ number of *doubly occupied* spatial MOs $\{\phi_i(\mathbf{r})\}$, $i = 1, 2, \ldots, n$

$m =$ number of basic spatial AOs $\{\chi_\mu(\mathbf{r})\}$, $\mu = 1, 2, \ldots, m$

$m \geq n \, m = n$ means *minimal* basis

$m > n$ usually includes *polarization* functions (e.g. 3d, 4f, 5g, ... on C, N, O, F atoms and 2p, 3d, 4f, ... on H atoms, as explained in Section 7.3).

We then see that the many-electron single determinant wavefunction Ψ has the following properties.

7.1.1 The Fock–Dirac Density Matrix

All physical properties of the system are determined by the *fundamental physical invariant* ρ, a one-electron bilinear function in the coordinates $\mathbf{x}$

and $\mathbf{x}'$, defined as

$$\rho(\mathbf{x}; \mathbf{x}') = \sum_{i=1}^{N} \psi_i(\mathbf{x})\psi_i^*(\mathbf{x}') \tag{7.3}$$

which is called the Fock–Dirac density matrix, and whose formal mathematical properties are

$$\int d\mathbf{x}\, \rho(\mathbf{x}; \mathbf{x}')|_{\mathbf{x}'=\mathbf{x}} = \int d\mathbf{x}\, \rho(\mathbf{x}; \mathbf{x}) = N \qquad \text{conservation} \tag{7.4}$$

$$\int d\mathbf{x}''\, \rho(\mathbf{x}; \mathbf{x}'')\rho(\mathbf{x}''; \mathbf{x}') = \rho(\mathbf{x}; \mathbf{x}') \qquad \text{idempotency} \tag{7.5}$$

$$\rho'(\mathbf{x}; \mathbf{x}') = \rho(\mathbf{x}; \mathbf{x}') \quad \text{invariance} \tag{7.6}$$

where ρ' is the result of any unitary transformation among its spin-orbitals.

Mathematically speaking, $\rho(\mathbf{x}; \mathbf{x}')$ is the kernel of the *integral operator*

$$\hat{\rho}(\mathbf{x}) = \int d\mathbf{x}'\, \rho(\mathbf{x}; \mathbf{x}')\hat{P}_{\mathbf{x}\mathbf{x}'} \tag{7.7}$$

which transforms a function $F(\mathbf{x})$ into

$$\hat{\rho}(\mathbf{x})F(\mathbf{x}) = \int d\mathbf{x}'\, \rho(\mathbf{x}; \mathbf{x}')\hat{P}_{\mathbf{x}\mathbf{x}'}F(\mathbf{x}) = \int d\mathbf{x}'\, \rho(\mathbf{x}; \mathbf{x}')F(\mathbf{x}') \tag{7.8}$$

a *new* function of $\mathbf{x}$, namely

$$\hat{\rho}F = \int d\mathbf{x}'\, \rho(\mathbf{x}; \mathbf{x}')F(\mathbf{x}') = \int d\mathbf{x}' \sum_{i=1}^{N} \psi_i(x)\psi_i^*(\mathbf{x}')F(\mathbf{x}') = \sum_{i=1}^{N} |\psi_i\rangle\langle\psi_i|F\rangle$$

$$\tag{7.9}$$

where we introduced the Dirac notation for the last term. In other words, $\hat{\rho}$ is a *projection operator* in Fock space which projects out of the arbitrary regular function $F(\mathbf{x})$ its N-components along the basic vectors $|\psi_i\rangle$, $i = 1, 2, \ldots, N$.

In a short notation, properties (7.4)–(7.6) can be symbolically written as

$$\mathrm{tr}\rho = N, \quad \rho^2 = \rho, \quad \rho' = \rho \tag{7.10}$$

It is often said that Equation 7.3 provides a spin-orbital *representation* of the projector $\hat{\rho}$.

7.1.2 Electronic Energy Expression

The general expression for the electronic energy of the many-electron system was given in Chapter 6 as

$$E_e = \int d\mathbf{x}_1 \, \hat{h}_1 \rho_1(\mathbf{x}_1; \mathbf{x}'_1)|_{\mathbf{x}'_1 = \mathbf{x}_1} + \frac{1}{2} \int d\mathbf{x}_1 \, d\mathbf{x}_2 \, \frac{1}{r_{12}} \rho_2(\mathbf{x}_1 \mathbf{x}_2; \mathbf{x}_1 \mathbf{x}_2) \tag{7.11}$$

But, Lennard-Jones (1931) showed that in HF theory

$$\rho_1(\mathbf{x}; \mathbf{x}') = \rho(\mathbf{x}; \mathbf{x}') \tag{7.12}$$

$$\rho_2(\mathbf{x}_1 \mathbf{x}_2; \mathbf{x}'_1 \mathbf{x}'_2) = \begin{vmatrix} \rho(\mathbf{x}_1; \mathbf{x}'_1) & \rho(\mathbf{x}_1; \mathbf{x}'_2) \\ \rho(\mathbf{x}_2; \mathbf{x}'_1) & \rho(\mathbf{x}_2; \mathbf{x}'_2) \end{vmatrix}$$
$$= \rho(\mathbf{x}_1; \mathbf{x}'_1)\rho(\mathbf{x}_2; \mathbf{x}'_2) - \rho(\mathbf{x}_1; \mathbf{x}'_2)\rho(\mathbf{x}_2; \mathbf{x}'_1) \tag{7.13}$$

Hence, in HF theory, the electronic energy assumes the characteristic one-electron form (*no* electron correlation $\Rightarrow$ IPM):

$$E_e = \int d\mathbf{x} \, \hat{h}\rho(\mathbf{x}; \mathbf{x}')|_{\mathbf{x}'=\mathbf{x}} + \frac{1}{2} \int d\mathbf{x} \, [J(\mathbf{x}) - \hat{K}(\mathbf{x})] \, \rho(\mathbf{x}; \mathbf{x}')|_{\mathbf{x}'=\mathbf{x}} \tag{7.14}$$

where both one-electron and two-electron components of E_e are expressed in terms of the fundamental invariant ρ and of Coulomb and exchange one-electron potentials:

$$J(\mathbf{x}_1) = \int d\mathbf{x}_2 \, \frac{\rho(\mathbf{x}_2; \mathbf{x}_2)}{r_{12}} \tag{7.15}$$

$$\hat{K}(\mathbf{x_1}) = \int d\mathbf{x_2}\, \frac{\rho(\mathbf{x_1};\mathbf{x_2})}{r_{12}} \hat{P}_{\mathbf{x_1}\mathbf{x_2}} \qquad (7.16)$$

The exchange potential $\hat{K}$ is an integral operator with kernel $\rho(\mathbf{x_1};\mathbf{x_2})/r_{12}$ and J is a multiplicative operator that involves the diagonal element of the Fock–Dirac density matrix.

In HF theory it is customary to introduce the one-electron Fock operator

$$\hat{F}(\mathbf{x}) = \hat{h} + J(\mathbf{x}) - \hat{K}(\mathbf{x}) \qquad (7.17)$$

where

$$\hat{h} = -\frac{1}{2}\nabla^2 + V, \quad V = -\sum_{\alpha} \frac{Z_\alpha}{r_\alpha} \qquad (7.18)$$

Therefore:

$$\hat{F}(\mathbf{x}) = -\frac{1}{2}\nabla^2 + V_{\text{eff}}, \quad V_{\text{eff}} = V + J - \hat{K} \qquad (7.19)$$

The Fock operator is the sum of the kinetic energy operator and the *effective* potential V_{eff} felt by each electron, due to the nuclear attraction and the Coulomb-exchange field of *all* electrons. In terms of the Fock operator (7.17), it is seen that

$$E_e[\rho] = \int d\mathbf{x}\, \hat{F}(\mathbf{x})\rho(\mathbf{x};\mathbf{x'})|_{\mathbf{x'}=\mathbf{x}} - \frac{1}{2}\int d\mathbf{x}\, [J(\mathbf{x}) - \hat{K}(\mathbf{x})]\rho(\mathbf{x};\mathbf{x'})|_{\mathbf{x'}=\mathbf{x}} \qquad (7.20)$$

where the average electron repulsion must now be *subtracted* in order to avoid counting it twice.

Hall (1951) and Roothaan (1951b) have shown independently that the *best* variational energy[1] for the HF determinant (7.2) is obtained when the spin-orbitals satisfy the HF *equations*:

$$\hat{F}(\mathbf{x})\psi_i(\mathbf{x}) = \varepsilon_i\psi_i(\mathbf{x}) \quad i = 1, 2, \ldots, N \qquad (7.21)$$

---

[1]Subject to the constraint of keeping orthonormal the spin-orbitals $\Rightarrow$ method of Lagrange multipliers as applied to the energy functional (7.20).

where $\hat{F}$ is the *same* for all electrons. Despite their simple aspect, (7.21) are complicated *integro-differential*[2] equations in which the operator $\hat{F}$ depends on the $\{\psi_i\}$ for which the equation should be solved, so that they must be solved by *iteration* $\Rightarrow$ hence the SCF method starting from any convenient initial guess.[3] The iteration must be stopped when the spin-orbitals obtained as solutions of the HF equations do not differ appreciably from those used in the construction of $\hat{F}(\rho) \Rightarrow \hat{F}\{\psi_i\}$. This is usually done by putting a convenient threshold on the energies.

7.2   ROOTHAAN FORMULATION OF THE LCAO–MO–SCF EQUATIONS

Eliminating spin from $\hat{F}(\mathbf{x}) = \hat{F}(\mathbf{rs})$, we obtain the *spinless* Fock operator:

$$\hat{F}(\mathbf{r}) = \hat{h} + 2J(\mathbf{r}) - \hat{K}(\mathbf{r}) \tag{7.22}$$

where

$$J(\mathbf{r}_1) = \int d\mathbf{r}_2 \, \frac{R(\mathbf{r}_2; \mathbf{r}_2)}{r_{12}} \tag{7.23}$$

$$\hat{K}(\mathbf{r}_1) = \int d\mathbf{r}_2 \, \frac{R(\mathbf{r}_1; \mathbf{r}_2)}{r_{12}} \hat{P}_{\mathbf{r}_1 \mathbf{r}_2} \tag{7.24}$$

are spinless Coulomb and exchange potentials (space only), and

$$R(\mathbf{r}_1; \mathbf{r}_2) = \rho^\alpha(\mathbf{r}_1; \mathbf{r}_2) = \rho^\beta(\mathbf{r}_1; \mathbf{r}_2) = \sum_i^{occ} \phi_i(\mathbf{r}_1)\phi_i^*(\mathbf{r}_2) \tag{7.25}$$

is the Fock–Dirac density matrix for closed shells, the summation in the last term of (7.25) being over all *occupied* orbitals $(i = 1, 2, \ldots, n)$.

---

[2] $\hat{K}$ is the *integral* operator (7.16) while ∇^2 is a *differential* operator.
[3] Usually, Hückel orbitals.

The HF equations for the occupied spatial MOs hence become

$$\hat{F}(\mathbf{r})\phi_i(\mathbf{r}) = \varepsilon_i\phi_i(\mathbf{r}) \quad i = 1, 2, \ldots, n = \frac{N}{2} \tag{7.26}$$

R has the projection properties of a density matrix in the space of occupied MOs:

$$\mathrm{tr}R = n, \quad R^2 = R, \quad R' = R \tag{7.27}$$

Introducing a basis of $m \, (\geq n)$ AOs, if $\mathbf{C}$ is the rectangular $(m \times n)$ matrix of the LCAO coefficients

$$\phi_i(\mathbf{r}) = \sum_{\mu=1}^{m}\chi_\mu(\mathbf{r})c_{\mu i}, \quad \phi(\mathbf{r}) = \chi(\mathbf{r})\mathbf{C} \tag{7.28}$$

then the Roothaan pseudoeigenvalue equations for the ith MO are best formulated in matrix form as

$$\mathbf{F}\mathbf{c}_i = \varepsilon_i\mathbf{M}\mathbf{c}_i \quad i = 1, 2, \ldots, n \tag{7.29}$$

and for the *whole* set of the n occupied MOs

$$\mathbf{F}\mathbf{C} = \mathbf{M}\mathbf{C}\varepsilon \tag{7.30}$$

where

$$\mathbf{F} = \chi^\dagger\hat{F}\chi \tag{7.31}$$

is the $(m \times m)$ matrix representative of the spinless Fock operator over the m basis functions $\chi = (\chi_1\chi_2\cdots\chi_m)$

$$\mathbf{M} = \chi^\dagger\chi \tag{7.32}$$

is the $(m \times m)$ metric matrix of the atomic basis, and

$$\varepsilon = \begin{pmatrix} \varepsilon_1 & 0 & \cdots & 0 \\ 0 & \varepsilon_2 & \cdots & 0 \\ \cdots & \cdots & \cdots & \cdots \\ 0 & 0 & \cdots & \varepsilon_n \end{pmatrix} \tag{7.33}$$

is the $(n \times n)$ diagonal matrix of the eigenvalues pertaining to occupied MOs.

The density matrix R in the atomic basis will be

$$R(\mathbf{r}; \mathbf{r}') = \phi(\mathbf{r})\phi(\mathbf{r}')^{\dagger} = \chi(\mathbf{r})\mathbf{C}\mathbf{C}^{\dagger}\chi(\mathbf{r}')^{\dagger} = \chi(\mathbf{r})\mathbf{R}\chi(\mathbf{r}')^{\dagger}, \quad \mathbf{R} = \mathbf{C}\mathbf{C}^{\dagger}$$
$$(7.34)$$

where $\mathbf{R}$ is the $(m \times m)$ matrix representative of the Fock–Dirac density matrix over the AO basis χ (hence follows the analysis of the electron distribution in the molecule $\Rightarrow$ Mulliken population analysis, a generalization of the simple example given in Section 6.3). The projection operator properties of matrix $\mathbf{R}$ in the AO basis can be written as

$$\mathrm{tr}\mathbf{R}\mathbf{M} = n, \quad \mathbf{R}\mathbf{M}\mathbf{R} = \mathbf{R} \qquad (7.35)$$

Roothaan's equations (7.29) are solved *iteratively* from the pseudosecular equation

$$|\mathbf{F} - \varepsilon\mathbf{M}| = 0 \qquad (7.36)$$

which is an algebraic equation in ε having as m roots the MO orbital energies

$$\underbrace{\varepsilon_1, \varepsilon_2, \ldots, \varepsilon_n}_{\text{occ}} \Big| \underbrace{\varepsilon_{n+1}, \varepsilon_{n+2}, \ldots, \varepsilon_m}_{\text{unocc}} \qquad (7.37)$$

illustrated in Figure 7.1.

According to Hund's rule, for nondegenerate levels, in the molecular ground state the first n levels are occupied by electrons with opposite spin (*bonding* levels, $\varepsilon_i < 0$) and the remaining $(m - n)$ levels are unoccupied (empty, *antibonding* levels, $\varepsilon_i > 0$). The highest occupied MO (HOMO) is ϕ_n (orbital energy ε_n) and the lowest unoccupied MO (LUMO) is ϕ_{n+1} (orbital energy ε_{n+1}).

The Roothaan SCF electronic energy is given in matrix form as

$$E_e = 2 \int d\mathbf{r} \, \hat{F}(\mathbf{r}) R(\mathbf{r}; \mathbf{r}')|_{\mathbf{r}'=\mathbf{r}} - \int d\mathbf{r} \, \hat{G}(\mathbf{r}) R(\mathbf{r}; \mathbf{r}')|_{\mathbf{r}'=\mathbf{r}}$$
$$= 2\mathrm{tr}\mathbf{F}\mathbf{R} - \mathrm{tr}\mathbf{G}\mathbf{R} \qquad (7.38)$$

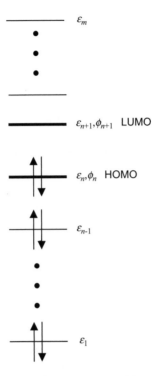

ε_m

$\varepsilon_{n+1}, \phi_{n+1}$ LUMO

ε_n, ϕ_n HOMO

ε_{n-1}

ε_1

Figure 7.1   Diagram of the orbital energies in an SCF calculation

where

$$G(\mathbf{r}) = 2J(\mathbf{r}) - \hat{K}(\mathbf{r}) \qquad (7.39)$$

is called the Roothaan matrix of total electron interaction. The matrix elements of (7.38) over AOs are

$$R_{\mu\nu} = \sum_{i}^{\text{occ}} c_{\mu i} c_{\nu i}^* \qquad (7.40)$$

$$F_{\mu\nu} = h_{\mu\nu} + G_{\mu\nu}, \quad h_{\mu\nu} = \langle \chi_\mu | \hat{h} | \chi_\nu \rangle \qquad (7.41)$$

$$G_{\mu\nu} = \sum_{\lambda} \sum_{\sigma} 2R_{\lambda\sigma} \left[(\chi_\nu \chi_\mu | \chi_\lambda \chi_\sigma) - \frac{1}{2} (\chi_\lambda \chi_\mu | \chi_\nu \chi_\sigma) \right] \qquad (7.42)$$

where the two-electron integrals are written in the charge density notation:

$$(\chi_\nu\chi_\mu|\chi_\lambda\chi_\sigma) = \int dr_1\, dr_2 \frac{\chi_\lambda(r_2)\chi_\sigma^*(r_2)}{r_{12}}\chi_\nu(r_1)\chi_\mu^*(r_1) \qquad (7.43)$$

Under appropriate simplifying assumptions in the matrix elements above, all LCAO approximations that can be derived from HF theory, such as Hückel, extended Hückel, Pople's complete neglect of differential overlap (CNDO), intermediate neglect of differential overlap (INDO), and so on, are easily obtained.

7.3   MOLECULAR SELF-CONSISTENT-FIELD CALCULATIONS

The AOs used as a basis in quantum chemical calculations are the STOs and GTOs introduced in Chapter 3. Calculations are affected by errors arising (i) from the very nature of the basis set (we saw that STOs are definitely superior to GTOs) and (ii) from the insufficiency of the basis set, the so-called truncation errors. We now briefly review the different types of bases most used in today molecular calculations.

A *minimal* basis set involves those AOs which are occupied in the ground state of the constituent atoms. An *extended* set includes polarization functions, namely those AOs that are unoccupied in the ground state of the atoms, say 3d and 4f AOs for C, N, O, F and 2p and 3d AOs for H.

In general, we speak of single zeta (SZ) for a minimal set, double zeta (DZ) when each AO is described by two functions, triple zeta (TZ) when each AO is described by three functions, and so on. When polarization functions (P) are included, we have correspondingly DZP, TZP or DZPP, TZPP (Van Duijneveldt-Van de Rijdt and Van Duijneveldt, 1982), depending on whether the polarization functions are only on the heavy atom or on the H atom as well.

GTO bases are often of the redundant Cartesian type:

$$G_{uvw}(r) = Nx^u y^v z^w \exp(-cr^2) \qquad (7.44)$$

whose number is given by the binomial coefficient

$$\binom{L+2}{L} = \frac{1}{2}(L+1)(L+2)$$

if $L = u + v + w$.

To reduce the large number of *primitives* usually needed in GTO calculations it is customary to resort to *contracted* GTOs, where each function is the sum of a certain number of primitives, each contraction scheme being specified by *fixed* numerical coefficients. The best basis is, of course, uncontracted.

As an example, a contraction scheme used for LiH by Tunega and Noga (1998) is based on the following *spherical* GTOs for Li and H:

$$(14s\, 8p\, 6d\, 5f | 12s\, 8p\, 6d\, 5f) \Rightarrow [11s\, 8p\, 6d\, 5f | 9s\, 8p\, 6d\, 5f] \qquad (7.45)$$

The 204 primitives (103 GTOs on Li, 101 on H) are contracted to 198 functions (100 GTOs on Li, 98 GTOs on H), with the polarization functions left uncontracted. This gives a rather moderate contraction.

Another example of a more sensible contraction can be taken from Lazzeretti and Zanasi (1981) in their roughly HF *Cartesian* GTO calculation on H_2O, which includes polarization functions on O and H:

$$(14s\, 8p\, 3d\, 1f | 10s\, 2p\, 1d) \Rightarrow [9s\, 6p\, 3d\, 1f | 6s\, 2p\, 1d] \qquad (7.46)$$

The 110 GTO primitives are here reduced to 91 contracted GTOs, giving an SCF molecular energy of $-76.066\,390E_h$. A more extended basis for H_2O was the $[13s\, 10p\, 5d\, 2f | 8s\, 4p\, 1d]$ contracted Cartesian GTO basis set recently used by Lazzeretti (personal communication, 2004) in an SCF calculation on H_2O, giving a molecular energy of $-76.066\,87E_h$, only $0.63 \times 10^{-3}E_h$ above the HF limit of $-76.067\,50E_h$ *estimated* by Rosenberg and Shavitt (1975) for H_2O.

Use is sometimes made of *even-tempered* (or *geometrical*) sequences of primitives, where the orbital exponents c_i are restricted by

$$c_i = ab^i \quad i = 1, 2, \ldots, m \qquad (7.47)$$

with a and b fixed and different for functions of s, p, d, f, ... symmetry. Thus, the number of nonlinear parameters (orbital exponents) to be optimized in a variational calculation is drastically reduced.

Table 7.1 Correlation-consistent polarized basis sets for first-row atoms B through to Ne

	Primitive	(sp) set contracted	Polarization set
cc-pVDZ	(9s4p)	[3s2p]	(1d)
cc-pVTZ	(10s5p)	[4s3p]	(2d1f)
cc-pVQZ	(12s6p)	[5s4p]	(3d2f1g)

Also widely used today are the GTO bases introduced by Pople and coworkers in the different versions of GAUSSIAN programmes: STO-nG, 6–31G, 6–31G*, 6–31G**, to denote, respectively, (i) n-GTOs to represent a single STO, (ii) six GTOs for the inner-shell + split-valence GTOs, three inner + one outer, (iii) the same with additional polarization functions on the heavy atom, and (iv) the same with additional polarization functions on the H atoms.

For use in calculations going beyond HF, Dunning (1989) introduced well-balanced correlation-consistent polarized valence GTO basis sets for the first-row atoms from B through to Ne (Table 7.1) and H (Table 7.2), denoted by the acronym cc-pVXZ, where X = D,T,Q,... (double, triple, quadruple, ...) indicates the number of functions in the original basis set, as shown below (Tables 7.1 and 7.2).

These bases were further elaborated by Woon and Dunning (1995) by extending the correlation-consistent polarized valence basis sets (cc-pVXZ) to include core-valence correlation effects giving (cc-pCVXZ) bases for the same atoms B through to Ne. For polarizabilities, additional *diffuse* functions were added, giving the so-called augmented basis sets aug-cc-pCVXZ. It must be admitted that such acronyms look rather esoteric to the uninitiated reader!

An *atomic* and a *molecular* example are given in Tables 7.3 and 7.4, where a comparison is made between different STO and GTO SCF results for ground-state Ne(^{1}S) and HF($^1\Sigma^+$) at the experimentally observed internuclear distance $R_e = 1.7328a_0$. The GTO bases in Table 7.3 are geometrical basis sets, whereas the GTO basis sets in Table 7.4 are of Cartesian type.

Table 7.2 Correlation-consistent polarized basis sets for H

	Primitive	(s) set contracted	Polarization set
cc-pVDZ	(5s)	[2s]	(1p)
cc-pVTZ	(6s)	[3s]	(2p1d)
cc-pVQZ	(7s)	[4s]	(3p2d1f)

Table 7.3 Comparison between STO,[a] GTO[b] and HF/2D[c] SCF results (atomic units) for the ground state of the Ne atom

STO basis/AO m	SZ 4	DZ 8	HF 23	HF/2D
E_e	− 127.812 2	− 128.535 11	− 128.547 05	− 128.547 13
ε_i/1s	− 32.662 13	− 32.759 88	− 32.772 48	− 32.772 454
2s	− 1.732 50	− 1.921 87	− 1.930 43	− 1.930 392
2p	− 0.561 72	− 0.841 43	− 0.850 44	− 0.850 411

GTO basis/AO m	Small 24	Intermediate 28	Large 37
E_e	− 128.464 724	− 128.528 123	− 128.543 969
ε_i/1s	− 32.761 82	− 32.757 62	− 32.765 52
2s	− 1.926 82	− 1.922 26	− 1.925 87
2p	− 0.841 96	− 0.842 87	− 0.846 30

[a] Clementi and Roetti (1974): SZ (1s1p), DZ (2s2p), HF (8s5p).
[b] Clementi and Corongiu (1982): small (9s5p), intermediate (10s6p), large (13s8p).
[c] Sundholm et al. (1985): (8s5p).

Table 7.4 Comparison between STO,[a–c] GTO[d] and HF/2D[e,f] SCF results (atomic units) for the ground state of the HF molecule at $R_e = 1.7328 a_0$

STO basis/AO m	SZ[a] 6	∼DZPP[b] 16	HF[c] 24[4]	HF/2D[e,f]
μ	0.584	0.852 95	0.764 0	0.756 076[f]
E	− 99.479	− 100.057 54	− 100.070 30	− 100.070 82[f]
ε_i/1σ	− 26.260 1	− 26.306 17	− 26.294 28	− 26.294 57[e]
2σ	− 1.484 9	− 1.610 68	− 1.600 74	− 1.600 99[e]
3σ	− 0.594 0	− 0.774 59	− 0.768 10	− 0.768 25[e]
1π	− 0.469 7	− 0.657 86	− 0.650 08	− 0.650 39[e]

GTO basis/AO m	STO-3G[d] 6	aug-ccpVDZ[d] 34	aug-ccpVTZ[d] 80
μ	0.506 95	0.759 83	0.756 19
E	− 98.570 775	− 100.034 422	− 100.061 868
ε_i/1σ	− 25.900 028	− 26.308 360	− 26.296 306
2σ	− 1.471 216	− 1.607 876	− 1.602 126
3σ	− 0.585 187	− 0.770 810	− 0.768 228
1π	− 0.464 165	− 0.650 201	− 0.650 247

[a] Ballinger (1959): SZ (2s3p1s).
[b] Clementi (1962): ∼DZPP (5s5p2d2s2p).
[c] Cade and Huo (1967): HF (5s8p3d2f3s2p1d).
[d] Lazzeretti and Pelloni (personal communication, 2006): STO-3G [2s1p1s], aug-ccpVDZ [4s3p2d3s2p], aug-ccpVTZ [5s4p3d2f4s3p2d].
[e] Sundholm (1985).
[f] Sundholm et al. (1985).

---

[4]In the top row of Table 7.4 the STOs were numbered individually since not all d and f functions on F or H were included in the calculation.

The tables show that the one-term approximation (SZ) is totally insufficient either for the atomic or the molecular case, giving large errors in total energy and orbital energies. The two-term (DZ) approximation improves both energies, but to reach the HF level a sensibly larger number of functions is needed. The comparison between STO and GTO results for Ne shows how much larger the number of GTOs must be to get an accuracy comparable to that of the corresponding STOs. Nevertheless, the nearly HF GTO basis for Ne, containing about twice as many functions as those of the corresponding STOs, is still in error by about $3 \times 10^{-3} E_h$ for the electronic energy and even more for the orbital energies.

In the molecular HF case (Table 7.4), it is seen that the SZ approximation is very poor, giving a molecular energy of about $0.592 E_h$ (371 kcal mol^{-1}!) above the HF limit of the last column and a charge distribution which severely *underestimates* the electric dipole moment μ. The nearly DZPP approximation of column 3 yields instead a sensible *overestimation* of the dipole moment. The nearly HF STO results of column 4 (Cade and Huo, 1967; also see McLean and Yoshimine (1967b)) are in fairly good agreement with the accurate HF/2D results (the 'benchmark') of the last column, based on the two-dimensional numerical quadrature of the one-electron HF equation, and, therefore, free from any basis set and truncation errors.

The same considerations are almost true for the GTO calculations of the bottom part of Table 7.4. STO-3G calculations are useless either for energy or dipole moment. The aug-ccpVDZ basis[5] overestimates dipole moment less than the corresponding STO set, but gives a worse energy. The aug-ccpVTZ basis[6] gives a good dipole moment, but still underestimates the energy, which was not unexpected since these Dunning sets are devised particularly for calculation of electric molecular properties.

7.4 HÜCKEL THEORY

Hückel theory amounts to a simple LCAO–MO theory of carbon π electrons in conjugated and aromatic hydrocarbons whose σ skeleton

---

[5] aug = one s, p, d *diffuse* functions on F, one s, p on H.

[6] aug = one s, p, d, f *diffuse* functions on F, one s, p, d on H.

is assumed planar.[7] Each carbon atom contributes an electron in its $2p\pi$ AO, AOs being assumed orthonormal and the coefficients determined by the Ritz method. The elements of the Hückel matrix $\mathbf{H}$ are given in terms of just two negative unspecified parameters, namely the diagonal α (the Coulomb integral) and the nearest-neighbour off-diagonal β (the resonance or bond integral), simply introduced in a *topological* way as

$$H_{\mu\mu} = \alpha \tag{7.48}$$

$$H_{\mu\nu} = \beta\delta_{\mu,\mu\pm1} \tag{7.49}$$

$$S_{\mu\nu} = \delta_{\mu\nu} \quad \mu,\nu = 1,2,\ldots,N \tag{7.50}$$

Hückel theory, therefore, is a noniterative MO theory distinguishing only between linear (open) or closed (ring) chains.

It is convenient to introduce the notation

$$\frac{\alpha - \varepsilon}{\beta} = -x, \quad \varepsilon = \alpha + x\beta \tag{7.51}$$

$$\Delta\varepsilon = \varepsilon - \alpha = x\beta \Rightarrow \frac{\Delta\varepsilon}{\beta} = x \tag{7.52}$$

so that x measures the π bond energy in units of β ($x > 0$ means *bonding*, $x < 0$ *antibonding*). The total π bond energy in Hückel theory[8] is then simply

$$\Delta E^{\pi} = \sum_{i}^{\mathrm{occ}} \Delta\varepsilon_i \tag{7.53}$$

while the π charge density in terms of the AOs is given by twice Equation 7.34 (Section 7.2).

Henceforth, we shall denote by D_N the Hückel determinant of order N, considering in detail the case of ethylene, the allyl radical, linear and

---

[7]σ-π separation is possible in this case.

[8]It is assumed that nuclear repulsion is essentially cancelled by the negative of the electron repulsion (Equation 7.20).

cyclic butadiene, hexatriene and benzene, taken as representatives of linear chains with $N = 2, 3, 4, 6$ and rings (closed chains) with $N = 4, 6$. For these systems, we shall solve in an elementary way for both eigenvalues and eigenvectors of the Hückel matrix $\mathbf{H}$, obtaining in this way the orbital energy levels and the schematic shapes of the MOs shown in Figures 7.2–7.7. General, and more powerful, techniques of solution are due to Lennard–Jones (1937) and will be used later in Section 7.5.

7.4.1   Ethylene ($N = 2$)

The Hückel secular equation for $N = 2$ is

$$D_2 = \begin{vmatrix} -x & 1 \\ 1 & -x \end{vmatrix} = x^2 - 1 = 0 \qquad (7.54)$$

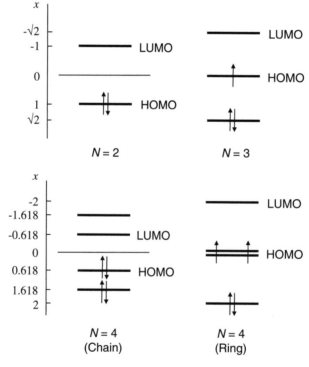

Figure 7.2   Hückel orbital energies for a linear chain with $N = 2, 3, 4$ and a ring with $N = 4$

whose roots (eigenvalues, left in the top row of Figure 7.2) are $x = \pm 1$, while the normalized eigenvectors (the MOs of the top row of Figure 7.3) are given by the (2×2) unitary matrix

$$
\mathbf{C} = (\mathbf{c}_1 \mathbf{c}_2) = \begin{pmatrix} \dfrac{1}{\sqrt{2}} & -\dfrac{1}{\sqrt{2}} \\[2mm] \dfrac{1}{\sqrt{2}} & \dfrac{1}{\sqrt{2}} \end{pmatrix} \tag{7.55}
$$

7.4.2 The Allyl Radical (N = 3)

The Hückel secular equation for $N = 3$ is

$$
D_3 = \begin{vmatrix} -x & 1 & 0 \\ 1 & -x & 1 \\ 0 & 1 & -x \end{vmatrix} = -x(x^2 - 1) + x = -x(x^2 - 2) = 0 \tag{7.56}
$$

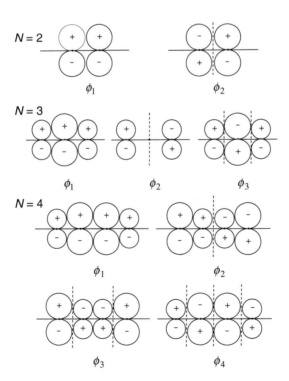

Figure 7.3 Hückel MOs for a linear chain with $N = 2, 3, 4$

whose roots (right in the top row of Figure 7.2) are $x = \pm\sqrt{2}, 0$, with the normalized eigenvectors (second row of Figure 7.3) given by the (3×3) unitary matrix

$$\mathbf{C} = (\mathbf{c_1 c_2 c_3}) = \begin{pmatrix} \dfrac{1}{2} & \dfrac{1}{\sqrt{2}} & \dfrac{1}{2} \\[3mm] \dfrac{\sqrt{2}}{2} & 0 & -\dfrac{\sqrt{2}}{2} \\[3mm] \dfrac{1}{2} & -\dfrac{1}{\sqrt{2}} & \dfrac{1}{2} \end{pmatrix} \qquad (7.57)$$

We now give below the explicit calculation of the eigenvectors corresponding to each eigenvalue. Hückel MOs (the eigenvectors) are obtained by solving in turn for each eigenvalue the linear homogeneous system associated with the secular Equation 7.56 with the additional constraint of coefficient normalization.

(a) $x_1 = \sqrt{2}$

$$\begin{cases} 1. & -\sqrt{2}c_1 + c_2 = 0 \\ 2. & c_1 - \sqrt{2}c_2 + c_3 = 0 \\ 3. & c_2 - \sqrt{2}c_3 = 0 \\ 4. & c_1^2 + c_2^2 + c_3^2 = 1 \end{cases} \qquad (7.58)$$

$\quad$ 1. $\quad$ gives: $\quad c_2 = \sqrt{2}c_1$

$\quad$ 3. $\quad$ gives: $\quad c_3 = \dfrac{1}{\sqrt{2}}c_2 = c_1$

so that it is immediately obtained that

$$c_1^2 + 2c_1^2 + c_1^2 = 4c_1^2 = 1 \Rightarrow c_1 = \frac{1}{2}, \quad c_2 = \frac{\sqrt{2}}{2}, \quad c_3 = \frac{1}{2} \qquad (7.59)$$

which is the *first* column of matrix (7.57).

(b) $x_2 = 0$

$$\begin{cases} c_2 = 0 \\ c_1 + c_3 = 0 \\ c_1^2 + c_2^2 + c_3^2 = 1 \end{cases} \qquad (7.60)$$

giving

$$c_2 = 0, \quad c_3 = -c_1, \quad c_1^2 + c_1^2 = 2c_1^2 \Rightarrow c_1 = \frac{1}{\sqrt{2}}, \quad c_2 = 0, \quad c_3 = -\frac{1}{\sqrt{2}}$$

$$(7.61)$$

which is the *second* column of matrix (7.57).

(c) $x_3 = -\sqrt{2}$

$$\begin{cases} 1. & \sqrt{2}c_1 + c_2 = 0 \\ 2. \ c_1 + \sqrt{2}c_2 + c_3 = 0 \\ 3. & c_2 + \sqrt{2}c_3 = 0 \\ & c_1^2 + c_2^2 + c_3^2 = 1 \end{cases} \qquad (7.62)$$

1. gives now : $\quad c_2 = -\sqrt{2}c_1$

3. gives : $\qquad c_3 = -\frac{1}{\sqrt{2}}c_2 = c_1$

so that it is immediately obtained that

$$c_1^2 + 2c_1^2 + c_1^2 = 4c_1^2 = 1 \Rightarrow c_1 = \frac{1}{2}, \quad c_2 = -\frac{\sqrt{2}}{2}, \quad c_3 = \frac{1}{2} \quad (7.63)$$

which is the *third* column of matrix (7.57).

It is of interest to evaluate charge and spin density distributions in the allyl radical, asssuming that the unpaired electron has α spin $(S = M_S = \frac{1}{2})$. It is convenient to write explicitly the three MOs as

$$\phi_1 = \frac{1}{2}(\chi_1 + \sqrt{2}\chi_2 + \chi_3), \quad \phi_2 = \frac{1}{\sqrt{2}}(\chi_1 - \chi_3), \quad \phi_3 = \frac{1}{2}(\chi_1 - \sqrt{2}\chi_2 + \chi_3)$$

$$(7.64)$$

We then have the following for the ρ^α and ρ^β components of the distribution function:

$$\rho^\alpha = \phi_1^2 + \phi_2^2 = \frac{1}{4}\chi_1^2 + \frac{2}{4}\chi_2^2 + \frac{1}{4}\chi_3^2 + \frac{2}{4}\chi_1^2 + \frac{2}{4}\chi_3^2 = \frac{3}{4}\chi_1^2 + \frac{2}{4}\chi_2^2 + \frac{3}{4}\chi_3^2$$

$$(7.65)$$

$$\rho^\beta = \phi_1^2 = \frac{1}{4}\chi_1^2 + \frac{2}{4}\chi_2^2 + \frac{1}{4}\chi_3^2 \tag{7.66}$$

This gives for the electron density

$$P(\mathbf{r}) = \rho^\alpha + \rho^\beta = \chi_1^2 + \chi_2^2 + \chi_3^2 \tag{7.67}$$

so that the charge distribution of the π electrons is *uniform* (one electron onto each carbon atom), as expected for an alternant hydrocarbon,[9] and for the spin density we have

$$Q(\mathbf{r}) = \rho^\alpha - \rho^\beta = \frac{1}{2}\chi_1^2 + \frac{1}{2}\chi_3^2 \tag{7.68}$$

According to (7.68), in the Hückel spin density the α unpaired electron is $\frac{1}{2}$ on atom 1 and $\frac{1}{2}$ on atom 3, being zero at the central atom. This is contrary to the ESR experimental observation that some α spin at the end atoms induces some β spin at the middle atom. This wrong result is due to the lack of any electron correlation in the wavefunction, which belongs to the class of IPM functions. Both (7.67) and (7.68) satisfy the appropriate conservation relations:[10] $\mathrm{tr}P = 3$ (the total number of π electrons) and $\mathrm{tr}Q = 1$ (the unpaired π electron of α spin).

The π bond energy in allyl is $2\sqrt{2} = 2.828$ (units of β), compared with 2 for the π bond energy of the ethylenic double bond (the prototype of the double bond). The difference

$$\Delta E^\pi(\text{allyl}) - \Delta E^\pi(\text{ethylene}) = 2.828 - 2 = 0.828 \tag{7.69}$$

is called the *delocalization energy* of the double bond in the allyl radical. It corresponds to a stabilization of the conjugated π system in the radical.

---

[9] Alternant hydrocarbons are conjugated molecules in which the carbon atoms can be divided into two sets, crossed and circled, such that no two members of the same set are bonded together. All molecules considered here are alternant hydrocarbons, in which energy levels occur in pairs, with a π bond energy $\pm x$, and coefficients of the paired MOs which are either the same or change sign (Murrell *et al.*, 1985).

[10] Equations 6.31 and 6.32.

7.4.3   Butadiene (N = 4)

The Hückel secular equation for the *linear* chain with $N = 4$ is

$$D_4 = \begin{vmatrix} -x & 1 & 0 & 0 \\ 1 & -x & 1 & 0 \\ 0 & 1 & -x & 1 \\ 0 & 0 & 1 & -x \end{vmatrix} = x^4 - 3x^2 + 1 = 0 \qquad (7.70)$$

where we have marked in boldface the top right and the bottom left elements that differ from those of the *closed* chain which will be examined in the next point. Equation 7.70 is a pseudoquartic equation that can be easily reduced to a quadratic equation by the substitution

$$x^2 = y \Rightarrow y^2 - 3y + 1 = 0 \qquad (7.71)$$

having the roots

$$y = \frac{3 \pm \sqrt{5}}{2} = \begin{cases} y_1 = \dfrac{3 + \sqrt{5}}{2} = 2.618 \\[2mm] y_2 = \dfrac{3 - \sqrt{5}}{2} = 0.382 \end{cases} \qquad (7.72)$$

So, we obtain the four roots (left in the bottom row of Figure 7.2)

$$x_1 = \sqrt{y_1} = 1.618, \quad \underset{\text{HOMO}}{x_2 = \sqrt{y_2} = 0.618,}$$

$$x_3 = -\sqrt{y_2} = -0.618, \quad x_4 = -\sqrt{y_1} = -1.618 \qquad (7.73)$$

$$\underset{\text{LUMO}}{}$$

with the first two being *bonding* levels and the last two *antibonding* levels.

The calculation of the MO coefficients (the eigenvectors corresponding to the four roots above) proceeds as we did for allyl, and the MOs (third and fourth row of Figure 7.3) are

$$\begin{cases} \phi_1 = 0.371\chi_1 + 0.601(\chi_2 + \chi_3) + 0.371\chi_4 \\ \phi_2 = 0.601\chi_1 + 0.371(\chi_2 - \chi_3) - 0.601\chi_4 \\ \phi_3 = 0.601\chi_1 - 0.371(\chi_2 + \chi_3) + 0.601\chi_4 \\ \phi_4 = 0.371\chi_1 - 0.601\chi_2 + 0.601\chi_3 - 0.371\chi_4 \end{cases} \qquad (7.74)$$

with the first two being *bonding* MOs (ϕ_2 = HOMO) and the last two *antibonding* MOs (ϕ_3 = LUMO).

Proceeding as we did for allyl, it is easily seen that the electron charge distribution is uniform (one π electron onto each carbon atom, alternant hydrocarbon) and the spin density is zero, as expected for a state with $S = M_S = 0$, since the two bonding MOs are fully occupied by electrons with opposite spin. The delocalization energy for linear butadiene is

$$\Delta E^\pi(\text{butadiene}) - 2\Delta E^\pi(\text{ethylene}) = 4.472 - 4 = 0.472 \qquad (7.75)$$

and, therefore, is sensibly less than the conjugation energy of the allyl radical.

7.4.4   Cyclobutadiene (N = 4)

The Hückel secular equation for the square *ring* with $N = 4$ is

$$D_4 = \begin{vmatrix} -x & 1 & 0 & \mathbf{1} \\ 1 & -x & 1 & 0 \\ 0 & 1 & -x & 1 \\ \mathbf{1} & 0 & 1 & -x \end{vmatrix} = x^4 - 4x^2 = x^2(x^2 - 4) = 0 \qquad (7.76)$$

where the boldface elements are the only ones differing from those of the *linear* chain (1 and 4 are now adjacent atoms). The roots of Equation 7.76 are $x_1 = 2$, $x_2 = x_3 = 0$ (doubly degenerate), and $x_4 = -2$ (right in the bottom row of Figure 7.2).

It is of interest to solve in some detail the homogeneous system originating (7.76) because of the presence of degenerate eigenvalues.

(a) $x_1 = 2$ The homogeneous system to be solved is

$$\begin{cases} 1. & -2c_1 + c_2 + c_4 = 0 \\ 2. & c_1 - 2c_2 + c_3 = 0 \\ 3. & c_2 - 2c_3 + c_4 = 0 \\ 4. & c_1 + c_3 - 2c_4 = 0 \\ & c_1^2 + c_2^2 + c_3^2 + c_4^2 = 1 \end{cases} \qquad (7.77)$$

Subtracting 4. from 2. gives $-2c_2 + 2c_4 = 0 \Rightarrow c_4 = c_2$.
    Then

$$1. \quad \text{gives:} \quad -2c_1 + 2c_2 = 0 \Rightarrow c_2 = c_1$$
$$2. \quad \text{gives:} \quad -c_1 + c_3 = 0 \Rightarrow c_3 = c_1$$
$$4c_1^2 = 1 \Rightarrow c_1 = c_2 = c_3 = c_4 = \frac{1}{2}$$

so that we finally obtain

$$\phi_1 = \frac{1}{2}(\chi_1 + \chi_2 + \chi_3 + \chi_4) \qquad (7.78)$$

the deepest *bonding* MO, sketched in the diagram of the top row of Figure 7.4. To simplify the drawings, the $2p\pi$ AOs in the MOs are viewed from above the zx molecular plane, so that only the signs of the upper lobes are reported.
(b) $x_2 = x_3 = 0$ (twofold degenerate eigenvalue)
The system is

$$\begin{cases} 1. & c_2 + c_4 = 0 \Rightarrow c_4 = -c_2 \\ 2. & c_1 + c_3 = 0 \Rightarrow c_3 = -c_1 \\ & c_1^2 + c_2^2 + c_3^2 + c_4^2 = 1 \end{cases} \qquad (7.79)$$

Equations 3 and 4 are useless, since they give the *same* result. Therefore, we have three equations and four unknowns, and there is one degree of freedom left, which means that one unknown can be chosen in an arbitrary way.

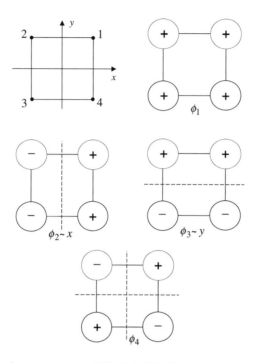

Figure 7.4   Coordinate system and Hückel MOs for cyclobutadiene

With reference to the coordinate system of Figure 7.4, we first choose $c_4 = c_1$ as additional arbitrary constraint. We then immediately obtain

$$\phi_2 = \phi_{2x} = \frac{1}{2}(\chi_1 - \chi_2 - \chi_3 + \chi_4) \propto x \qquad (7.80)$$

a function having the same transformation properties as the x-coordinate (left diagram in the second row of Figure 7.4). yz is a *nodal* plane for this MO. Next, we choose $c_4 = -c_1$ as additional arbitrary constraint, therefore obtaining

$$\phi_3 = \phi_{2y} = \frac{1}{2}(\chi_1 + \chi_2 - \chi_3 - \chi_4) \propto y \qquad (7.81)$$

a function having the same transformation properties as the y-coordinate (right diagram in the second row of Figure 7.4). zx is now a *nodal* plane for this MO. ϕ_{2x} and ϕ_{2y} are the pair of HOMOs belonging to the doubly degenerate energy level $\varepsilon_2 = \varepsilon_3$, which transform as the pair of basic vectors e_x and e_y of the symmetry D_{2h} (see Chapter 12) to which the σ skeleton of cyclobutadiene belongs. They are, therefore, orthogonal and not interacting, as can be immediately seen:

$$\langle \phi_{2x} | \phi_{2y} \rangle = \frac{1}{4} \langle \chi_1 - \chi_2 - \chi_3 + \chi_4 | \chi_1 + \chi_2 - \chi_3 - \chi_4 \rangle$$

$$= \frac{1}{4}(1 - 1 + 1 - 1) = 0 \tag{7.82}$$

$$\langle \phi_{2x} | H | \phi_{2y} \rangle = \frac{1}{4} \langle \chi_1 - \chi_2 - \chi_3 + \chi_4 | H | \chi_1 + \chi_2 - \chi_3 - \chi_4 \rangle$$

$$= \frac{1}{4}[2(\alpha + \beta - \beta) - 2(\alpha + \beta - \beta)] = 0 \tag{7.83}$$

It is also seen that they belong to the *same* eigenvalue $\varepsilon = \alpha$:

$$\varepsilon_{2x} = \frac{1}{4} \langle \chi_1 - \chi_2 - \chi_3 + \chi_4 | H | \chi_1 - \chi_2 - \chi_3 + \chi_4 \rangle$$

$$= \frac{1}{4}(4\alpha - 2\beta + 2\beta) = \alpha \tag{7.84}$$

$$\varepsilon_{2y} = \frac{1}{4} \langle \chi_1 + \chi_2 - \chi_3 - \chi_4 | H | \chi_1 + \chi_2 - \chi_3 - \chi_4 \rangle = \frac{1}{4}(4\alpha + 2\beta - 2\beta) = \alpha \tag{7.85}$$

(c) $x_4 = -2$

Now, the solution of the homogeneous system

$$\begin{cases} 1. & 2c_1 + c_2 + c_4 = 0 \\ 2. & c_1 + 2c_2 + c_3 = 0 \\ 3. & c_2 + 2c_3 + c_4 = 0 \\ 4. & c_1 + c_3 + 2c_4 = 0 \\ & c_1^2 + c_2^2 + c_3^2 + c_4^2 = 1 \end{cases} \tag{7.86}$$

does not present any problem. The calculation proceeds exactly in the same way as we did for the first eigenvalue, and the last *antibonding* MO (LUMO) is found to be

$$\phi_4 = \frac{1}{2}(\chi_1 - \chi_2 + \chi_3 - \chi_4) \tag{7.87}$$

ϕ_4 is sketched in the diagram of the bottom row of Figure 7.4 and shows the existence of two nodal planes orthogonal to each other.[11]

The delocalization energy for cyclobutadiene is

$$\Delta E^\pi(\text{cyclobutadiene}) - 2\Delta E^\pi(\text{ethylene}) = 4 - 4 = 0 \qquad (7.88)$$

so that, in Hückel theory, the π system of cyclobutadiene has *zero* delocalization energy, which is a rather unexpected result.

7.4.5 Hexatriene (N = 6)

The Hückel secular equation for the *linear* chain with $N = 6$ is

$$D_6 = \begin{vmatrix} -x & 1 & 0 & 0 & 0 & \mathbf{0} \\ 1 & -x & 1 & 0 & 0 & 0 \\ 0 & 1 & -x & 1 & 0 & 0 \\ 0 & 0 & 1 & -x & 1 & 0 \\ 0 & 0 & 0 & 1 & -x & 1 \\ \mathbf{0} & 0 & 0 & 0 & 1 & -x \end{vmatrix} = 0 \qquad (7.89)$$

where we have marked in boldface the top right and the bottom left elements that differ from those of the *closed* chain which corresponds to the benzene molecule. We obtain the sixth-degree equation

$$D_6 = x^6 - 5x^4 + 6x^2 - 1 = (x^3 + x^2 - 2x - 1)(x^3 - x^2 - 2x + 1) = 0 \qquad (7.90)$$

whose solutions are expected to occur in pairs of opposite sign, since only *even* powers of x appear in it. It has the following roots,[12] written in ascending order:

$$\begin{cases} x_1 = 1.802, & x_2 = 1.247, & x_3 = 0.445, \\ x_4 = -0.445, & x_5 = -1.247, & x_6 = -1.802 \end{cases} \qquad (7.91)$$

[11] As a general rule, the number of nodal planes increases for the higher π orbitals, while the deepest bonding MO has no nodal planes (except for the molecular plane, which is common to all molecules considered here).

[12] These cubic equations are not easily solved. The roots were evaluated using the Lennard-Jones formula (7.105).

These roots are symmetrical in pairs about $x = 0$, as it should be for an alternant hydrocarbon. In the factorized form of Equation 7.90, x_2, x_4, and x_6 are solutions of the first cubic equation and x_1, x_3, and x_5 of the second cubic equation.

The π energy levels of hexatriene are sketched in the left diagram of Figure 7.5. The delocalization energy for linear hexatriene is

$$\Delta E^{\pi}(\text{hexatriene}) - 3\Delta E^{\pi}(\text{ethylene}) = 2(1.802 + 1.247 + 0.445) - 6$$
$$= 6.988 - 6 = 0.988$$

$$(7.92)$$

We now turn to benzene, the closed chain corresponding to hexatriene.

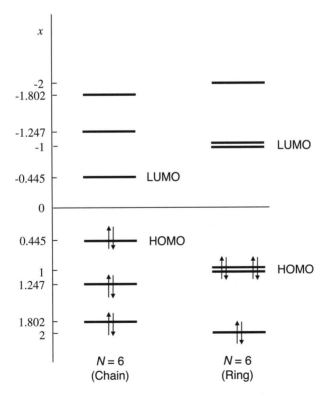

Figure 7.5 Hückel orbital energies for linear chain (hexatriene) and ring (benzene) for $N = 6$

7.4.6   Benzene ($N = 6$)

The Hückel secular equation for the *closed* chain (ring) with $N = 6$ is

$$D_6 = \begin{vmatrix} -x & 1 & 0 & 0 & 0 & \mathbf{1} \\ 1 & -x & 1 & 0 & 0 & 0 \\ 0 & 1 & -x & 1 & 0 & 0 \\ 0 & 0 & 1 & -x & 1 & 0 \\ 0 & 0 & 0 & 1 & -x & 1 \\ \mathbf{1} & 0 & 0 & 0 & 1 & -x \end{vmatrix} = 0 \qquad (7.93)$$

where the boldface elements are the two differing from those of the *linear* chain (1 and 6 are now adjacent atoms). By expanding the determinant we obtain a sixth-degree equation in x that can be easily factorized into the three quadratic equations:[13]

$$D_6 = x^6 - 6x^4 + 9x^2 - 4 = (x^2 - 4)(x^2 - 1)^2 = 0 \qquad (7.94)$$

with the following roots, written in ascending order:

$$x = 2, \quad 1, \quad 1, \quad -1, \quad -1, \quad -2 \qquad (7.95)$$

Because of the high symmetry of the molecule, two levels are now doubly degenerate. The calculation of the MO coefficients can be done using the elementary algebraic methods used previously for the allyl radical and cyclobutadiene. With reference to Figure 7.6, a rather lengthy

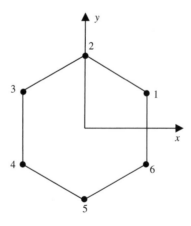

Figure 7.6 Numbering of carbon atoms in benzene and coordinate system

---

[13] The same factorization can be obtained using hexagonal symmetry. In hexatriene the symmetry is lower and only two cubic equations are obtained, equivalent to using the symmetry plane bisecting the molecule.

calculation shows that the *real* MOs are

$$
\begin{cases}
\phi_1 = \dfrac{1}{\sqrt{6}}(\chi_1 + \chi_2 + \chi_3 + \chi_4 + \chi_5 + \chi_6) \\[2mm]
\phi_2 = \dfrac{1}{2}(\chi_1 - \chi_3 - \chi_4 + \chi_6) \propto x \\[2mm]
\phi_3 = \dfrac{1}{\sqrt{12}}(\chi_1 + 2\chi_2 + \chi_3 - \chi_4 - 2\chi_5 - \chi_6) \propto y \\[2mm]
\phi_4 = \dfrac{1}{\sqrt{12}}(\chi_1 - 2\chi_2 + \chi_3 + \chi_4 - 2\chi_5 + \chi_6) \propto x^2 - y^2 \\[2mm]
\phi_5 = \dfrac{1}{2}(\chi_1 - \chi_3 + \chi_4 - \chi_6) \propto xy \\[2mm]
\phi_6 = \dfrac{1}{\sqrt{6}}(\chi_1 - \chi_2 + \chi_3 - \chi_4 + \chi_5 - \chi_6).
\end{cases}
\tag{7.96}
$$

The first degenerate MOs transform as (x,y) and are *bonding* (HOMOs); the second degenerate MOs transform as $(x^2 - y^2, xy)$ and are *antibonding* (LUMOs). These degenerate MOs have one and two nodal planes respectively. The highest antibonding MO, ϕ_6, has three nodal planes. The six real MOs for the π electron system of benzene are sketched in the drawings of Figure 7.7, where the same assumptions were made as for cyclobutadiene.

The ρ^α and ρ^β components of the distribution function for benzene are equal:

$$
\begin{aligned}
\rho^\alpha = \rho^\beta &= \phi_1^2 + \phi_2^2 + \phi_3^2 \\[2mm]
&= \chi_1^2\left(\frac{1}{6} + \frac{1}{4} + \frac{1}{12}\right) + \chi_2^2\left(\frac{1}{6} + \frac{4}{12}\right) + \chi_3^2\left(\frac{1}{6} + \frac{1}{4} + \frac{1}{12}\right) \\[2mm]
&\quad + \chi_4^2\left(\frac{1}{6} + \frac{1}{4} + \frac{1}{12}\right) + \chi_5^2\left(\frac{1}{6} + \frac{4}{12}\right) + \chi_6^2\left(\frac{1}{6} + \frac{1}{4} + \frac{1}{12}\right)
\end{aligned}
\tag{7.97}
$$

so that

$$
\rho^\alpha = \rho^\beta = \frac{1}{2}(\chi_1^2 + \chi_2^2 + \chi_3^2 + \chi_4^2 + \chi_5^2 + \chi_6^2)
\tag{7.98}
$$

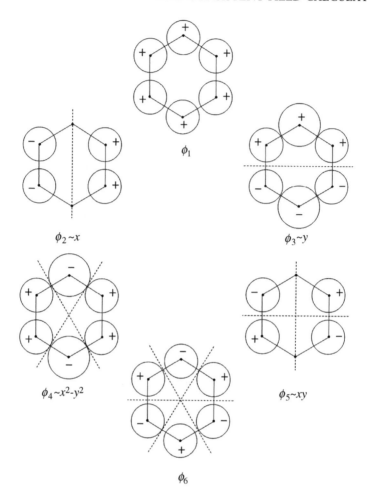

Figure 7.7 Hückel MOs for the π electrons in benzene (real form)

We then have

$$P(\mathbf{r}) = \rho^\alpha + \rho^\beta = \chi_1^2 + \chi_2^2 + \chi_3^2 + \chi_4^2 + \chi_5^2 + \chi_6^2 \qquad (7.99)$$

for the electron density and the charge distribution of the π electrons is *uniform* (one electron onto each carbon atom), as expected for an alternant hydrocarbon; whereas the spin density is zero:

$$Q(\mathbf{r}) = \rho^\alpha - \rho^\beta = 0 \qquad (7.100)$$

as it must be for a singlet state.

The π bond energy (units of β) for benzene is

$$\Delta E^{\pi} = 2 \times 2 + 4 \times 1 = 8 \qquad (7.101)$$

When the π bond energy of three ethylenes

$$3\Delta E^{\pi}(\text{ethylene}) = 3 \times 2 = 6 \qquad (7.102)$$

is subtracted from (7.101), we obtain the following for the delocalization energy of the three double bonds in benzene:

$$\Delta E^{\pi}(\text{benzene}) - 3\Delta E^{\pi}(\text{ethylene}) = 8 - 6 = 2 \qquad (7.103)$$

This is the highest value obtained so far among the molecules studied. This energy lowering is responsible for the great stability of the benzene molecule, where the three aromatic π bonds are fully delocalized and bear no resemblance at all to three ethylenic double bonds. Further stability in benzene arises from the fact that it has no strain in its σ skeleton.

It is also of interest to compare the π bond energy of benzene with that of hexatriene. We have

$$\Delta E^{\pi}(\text{benzene}) - \Delta E^{\pi}(\text{hexatriene}) = 8 - 6.988 = 1.012 \qquad (7.104)$$

so that ring *closure* in benzene is favoured, leading to further stabilization with respect to the open chain.

7.5   A MODEL FOR THE ONE-DIMENSIONAL CRYSTAL

Increasing the number of interacting AOs increases the number of resulting MOs. For the $C_N H_{N+2}$ polyene chain the MO levels, which always range between $\alpha + 2\beta$ and $\alpha - 2\beta$, become closer and closer up to transforming in *bands* (a continuous succession of molecular levels) which are characteristic of solids.

Using the general formula

$$\varepsilon_k = \alpha + 2\beta \cos \frac{\pi}{N+1} k \quad k = 1, 2, \ldots, N \qquad (7.105)$$

derived by Lennard-Jones (1937)[14] for the orbital energy of the kth MO in the N-atom linear polyene chain, the following results are easily established.

(i) First level ($k = 1$):

$$\varepsilon_1 = \alpha + 2\beta \cos \frac{\pi}{N+1} \quad \lim_{N \to \infty} \varepsilon_1 = \alpha + 2\beta \qquad (7.106)$$

Last level ($k = N$):

$$\varepsilon_N = \alpha + 2\beta \cos \frac{\pi N}{N+1} = \alpha + 2\beta \cos \frac{\pi}{1 + \frac{1}{N}} \quad \lim_{N \to \infty} \varepsilon_N = \alpha - 2\beta \qquad (7.107)$$

The difference between two successive levels is

$$\Delta \varepsilon = \varepsilon_{k+1} - \varepsilon_k = 2\beta \left[\cos \frac{\pi}{N+1}(k+1) - \cos \frac{\pi}{N+1} k \right]$$

$$= -4\beta \sin \frac{\pi}{2} \frac{2k+1}{N+1} \sin \frac{\pi}{2} \frac{1}{N+1} \qquad (7.108)$$

where use was made of the trigonometric identity

$$\cos \alpha - \cos \beta = -2 \sin \frac{\alpha + \beta}{2} \sin \frac{\alpha - \beta}{2} \qquad (7.109)$$

Hence, for $N \to \infty$, $\Delta \varepsilon \to 0$ and we have formation of a continuous band of molecular levels. The limiting values $\alpha + 2\beta$ and $\alpha - 2\beta$ are reached asymptotically when $N \to \infty$.

(ii) For $N \to \infty$, therefore, the polyene chain becomes the model for the one-dimensional crystal. We have a bonding band with energy ranging from $\alpha + 2\beta$ to α and an antibonding band with energy ranging from α to $\alpha - 2\beta$, which are separated by the so called Fermi level, the top of the bonding band occupied by electrons. It is important to note that using just one β, equal for single and double bonds, there is no *band gap* between bonding and antibonding levels.

---

[14]See also McWeeny (1979).

If we admit $|\beta_d| > |\beta_s|$, as reasonable and done by Lennard-Jones (1937) in his original study, then we have a band gap $\Delta = 2(\beta_d - \beta_s)$, which is of great importance in the properties of solids. Metals and covalent solids, conductors and insulators, and semiconductors can all be traced back to the model of the infinite polyene chain extended to three dimensions (McWeeny, 1979).

8

Post-Hartree–Fock Methods

In this chapter we shall briefly introduce some methods which, mostly starting from the uncorrelated HF approximation, attempt to reach chemical accuracy (1 kcal mol^{-1} or less) in the quantum chemical calculation of the atomization energies. We shall outline first the basic principles of configuration interaction (CI) and multiconfiguration SCF (MC-SCF) techniques, proceeding next to some applications of the so-called many-body perturbation methods, mostly the Møller–Plesset second-order approximation to the correlation energy (MP2), which is the starting point of the more efficient methods of accounting for correlation effects directly including the interelectronic distance in the wavefunction, such as the MP2-R12 and CC-R12 methods of the Kutzelnigg group. The chapter ends with a short introduction to density functional theory (DFT).

8.1 CONFIGURATION INTERACTION

Given a basis of atomic or molecular spin-orbitals, we construct a linear combination of electron configurations in the form of many-electron Slater determinants, with coefficients determined by the Ritz method, to give the CI wavefunction:

$$\Psi(\mathbf{x}_1, \mathbf{x}_2, \ldots, \mathbf{x}_N) = \sum_{\kappa} \Psi_{\kappa}(\mathbf{x}_1, \mathbf{x}_2, \ldots, x_N) C_{\kappa} \qquad (8.1)$$

When all possible configurations arising from a given basis set are included, we speak of the full-CI wavefunction. It should be recalled that

Methods of Molecular Quantum Mechanics: An Introduction to Electronic Molecular Structure
Valerio Magnasco
© 2009 John Wiley & Sons, Ltd

only configurations of given S and M_S belonging to a given molecular symmetry, have nonzero matrix elements of the molecular Hamiltonian. Even if the method is, in principle, exact if we include all configurations, expansion (8.1) converges usually quite slowly and the number of configurations becomes rapidly very large, involving up to millions of determinants.[1]

This is due to the difficulty of the wavefunction (8.1) in accounting for the *cusp condition* for each electron pair (Kutzelnigg, 1985):

$$\lim_{r_{ij} \to 0} \left(\frac{1}{\Psi_0} \frac{\partial \Psi}{\partial r_{ij}} \right) = \frac{1}{2} \tag{8.2}$$

which is needed to keep the wavefunction *finite* when $r_{ij} = 0$ in presence of the singularities of the Coulomb terms in the Hamiltonian.

Following Kato (1957), Kutzelnigg (1985) has shown that the Ψ expanded in powers of the interelectronic distance r_{ij}

$$\Psi = \Psi_0(1 + ar_{ij} + br_{ij}^2 + cr_{ij}^3 + \cdots) \tag{8.3}$$

satisfies this condition near the singular points for the pair of electrons i and j when $a = \frac{1}{2}$. In fact:

$$\frac{\partial \Psi}{\partial r_{ij}} = \Psi_0(a + 2b\, r_{ij} + \cdots) \tag{8.4}$$

$$\frac{1}{\Psi_0} \frac{\partial \Psi}{\partial r_{ij}} = a + 2b\, r_{ij} + \cdots \tag{8.5}$$

$$\lim_{r_{ij} \to 0} \left(\frac{1}{\Psi_0} \frac{\partial \Psi}{\partial r_{ij}} \right) = a = \frac{1}{2} \tag{8.6}$$

where a is a constant, with $a = \frac{1}{2}$ if i and j are both electrons, $a = -Z_B$ if i is a nucleus of charge $+Z_B$ and j an electron.

Using He as a simple example, Kutzelnigg (1985) showed that use of a starting wavefunction of the type

$$\Psi_0\left(1 + \frac{1}{2}r_{12}\right) \tag{8.7}$$

---

[1] Special techniques are required for solving the related secular equations of such huge dimensions (Roos, 1972).

where Ψ_0 is the simple product two-electron hydrogen-like wavefunction for ground-state He, gives a cusp-corrected CI expansion rapidly convergent with the biexcitations with $\ell = 0, 1, 2, 3, 4, \ldots (s^2, p^2, d^2, f^2, g^2, \ldots$ type): just 156 interconfigurational functions up to $\ell = 4$ give $E = -2.903\,722E_h$, roughly the same energy value obtained by including about 8000 interconfigurational functions with $\ell \geq 6$ in the ordinary CI expansion starting from Ψ_0. The accurate comparison value, due to Frankowski and Pekeris (1966), is $E = -2.903\,724\,377\,033E_h$, a 'benchmark' for the He atom correct to the last decimal figure (picohartree). Of course, use of the wavefunction (8.7) as a starting point in the CI expansion (8.1) involves the more difficult evaluation of unconventional one- and two-electron integrals.

8.2 MULTICONFIGURATION SELF-CONSISTENT-FIELD

In this method, mostly due to Wahl and coworkers (Wahl and Das, 1977), both the form of the orbitals in each single determinantal function and the coefficients of the linear combination of the configurations are optimized in a wavefunction like (8.1). The orbitals of a few valence-selected configurations are adjusted iteratively until self-consistency with the simultaneous optimization of the linear coefficients is obtained. The method predicts a reasonable well depth in He_2 and reasonable atomization energies (within 2 kcal mol^{-1}) for a few diatomics, such as H_2, Li_2, F_2, CH, NH, OH and FH.[2]

8.3 MØLLER–PLESSET THEORY

Since the Møller–Plesset approach is based on Rayleigh–Schroedinger (RS) perturbation theory, which will be introduced to some extent only in Chapter 10, it seems appropriate to give a short résumé of it here.

Stationary RS perturbation theory is based on the partition of the Hamiltonian $\hat{H}$ into an unperturbed Hamiltonian $\hat{H}_0$ and a *small* perturbation V, and on the expansion of the actual eigenfunction ψ and eigenvalue E into powers of the perturbation, each correction being specified by a definite *order* given by the power of an expansion parameter λ. For the method to be applied safely, it is necessary (i) that the expansion

---

[2] He_2, F_2 and NH are *not* bonded at the SCF level.

converges and (ii) that the unperturbed eigenfunction ψ_0 satisfies *exactly* the zeroth-order equation with eigenvalue E_0. While E_1, the first-order correction to the energy, is the average value of the perturbation over the unperturbed eigenfunction ψ_0 (a *diagonal* term), the second-order term E_2 is given as a transition (*nondiagonal*) integral in which state ψ_0 is changed into state ψ_1 under the action of the perturbation V. Further details are left to Chapter 10.

Møller–Plesset theory (Møller and Plesset, 1934) starts from $E(HF)$ considered as the result in first order of perturbation theory, $E(HF) = E_0 + E_1$, assuming as unperturbed ψ_0 the single determinant HF wavefunction, and as first-order perturbation the difference between the instantaneous electron repulsion and its average value calculated at the HF level. Therefore, it gives *directly* a second-order approximation to the correlation energy, since by definition

$$E(\text{correlation}) = E(\text{true}) - E(\text{HF}) \tag{8.8}$$

when $E(\text{true})$ is replaced by its second-order approximation

$$E(\text{true}) \approx E_0 + E_1 + E_2 = E(\text{HF}) + E_2(\text{MP}) \tag{8.9}$$

Hence:

$$E(\text{true}) - E(\text{HF}) \approx E_2(\text{MP}) = E(\text{MP2})$$
$$= \text{second-order approximation to the correlation energy}$$
$$\tag{8.10}$$

It is seen that only *biexcitations* can contribute to E_2, since monoexcitations give a zero contribution for HF Ψ_0 (Brillouin's theorem).

Comparison of SCF and MP2 results for the $^1A_1$ ground state of the H_2O molecule (Rosenberg *et al.*, 1976; Bartlett *et al.*, 1979) shows that MP2 improves greatly the properties (molecular geometry, force constants, electric dipole moment) but gives no more than 76% of the estimated correlation energy.

8.4 THE MP2-R12 METHOD

This is a Møller–Plesset second-order theory, devised by Kutzelnigg and coworkers (Klopper and Kutzelnigg, 1991),[3] which incorporates the

---

[3] Presented at the VIIth International Symposium on Quantum Chemistry, Menton, France, 2–5 July 1991.

linear r_{12}-dependent term into MP2 and MP3. Difficulties with the new three-electron integrals occurring in the theory are overcome in terms of expansions over ordinary two-electron integrals for nearly saturated basis sets. Fairly good results, improving upon ordinary MP2, are obtained for the correlation energies in simple closed-shell atomic and molecular systems (H_2, CH_4, NH_3, H_2O, HF, Ne) using extended sets of GTOs.

8.5 THE CC-R12 METHOD

The coupled-cluster (CC) method is a natural infinite-order generalization of many-body perturbation theory (MBPT), of which Møller–Plesset MP2 was the second-order approximation. In MBPT, starting from a reference wavefunction Ψ_0, multiple excitations from unperturbed occupied (occ) orbitals to unoccupied (empty) ones are considered. The theory was developed for use in many-body physics mostly in terms of rather awkward[4] second quantization and diagrammatic techniques (McWeeny, 1989).

In the CC method, the exact wavefunction is expressed in terms of an exponential form of the variational wavefunction,[5] where a *cluster operator* $\exp(\hat{T})$ acts upon a single-determinant reference wavefunction Ψ_0.[6] In the full CCSDT model, the cluster operator is usually truncated after $\hat{T}_3$ (triple excitations).

The CC-R12 method incorporates explicitly the interelectronic distance r_{12} into the wavefunction by replacing $\hat{T}$ by $\hat{S} = \hat{T} + \hat{R}$, where $\hat{R}$ is the r_{12}-contribution to the double excitation cluster operator. The operator $\hat{R}_{ij}^{kl}$ creates unconventionally substituted determinants in which a pair of occ orbitals i,j is replaced by another pair of occ orbitals l,k multiplied by the interelectronic distance r_{12}.

The CCSDT-R12 method devised by Noga and Kutzelnigg (1994) is the best available today for the computation of the atomization energies of simple molecules.[7] CCSDT-R12 calculations on ground-state NH_3, H_2O, FH, N_2, CO, F_2 at the experimentally observed geometries, using nearly saturated, well-balanced (spdfghlspdf) GTO basis sets, give atomization energies in perfect agreement with the experimental spectroscopic data

---

[4] At least for ordinary quantum chemists.

[5] German, ansatz.

[6] Possibly HF, in which case the contribution of monoexcitations vanishes because of Brillouin's theorem.

[7] It is actually (2007) in progress the extension of the theory to the calculation of second-order molecular properties, such as frequency-dependent polarizabilities.

(Noga *et al.*, 2001). It is hoped that in this way it will be possible to obtain 'benchmarks' in the calculation of atomization energies, at least for the small molecules of the first row.

8.6 DENSITY FUNCTIONAL THEORY

DFT was initially developed by Hohenberg and Kohn (1964) and by Kohn and Sham (1965), and is largely used today by the quantum chemical community in calculations on complex molecular systems. It must be stressed that DFT is a *semiempirical* theory accounting in part for electron correlation.

The electronic structure of the ground state of a system is assumed to be uniquely determined by the ground state electronic density $\rho_0(\mathbf{r})$, and a variational criterion is given for the determination of ρ_0 and E_0 from an arbitrary regular function $\rho(\mathbf{r})$. The variational optimization of the energy functional $E[\rho]$ constrained by the normalization condition:

$$E[\rho] \geq E[\rho_0], \quad \int d\mathbf{r}\, \rho(\mathbf{r}) = N \qquad (8.11)$$

shows that the functional derivative[8] is nothing but the effective one-electron Kohn–Sham Hamiltonian $\hat{h}^{KS}$:

$$\frac{\delta E[\rho]}{\delta \rho(\mathbf{r})} = -\frac{1}{2}\nabla^2 + V_{\text{eff}}(\mathbf{r}) = \hat{h}^{KS}(\mathbf{r}) \qquad (8.12)$$

where

$$V_{\text{eff}}(\mathbf{r}) = V(\mathbf{r}) + J(\mathbf{r}) + V_{\text{xc}}(\mathbf{r}) \qquad (8.13)$$

the effective potential at $\mathbf{r}$ is the sum of the electron–nuclear attraction potential V, plus the Coulomb potential J of the electrons of density ρ, plus the *exchange-correlation* potential V_{xc} for all the electrons. It is seen that the effective potential (8.13) differs from the usual HF potential by the undetermined *correlation* potential in V_{xc}. Since V_{xc} cannot be defined exactly, it can only be given semiempirical evaluations. Most used in applications is the Becke–Lee–Yang–Parr (B-LYP) correlation potential. Kohn–Sham orbitals $\phi_i^{KS}(\mathbf{r})$, $i = 1, 2, \ldots, n$, are then obtained from the

---

[8] The Euler–Lagrange parameter λ of the constrained minimization.

iterative SCF solution of the corresponding Kohn–Sham eigenvalue equations, much as we did for the HF equations of Chapter 7:

$$\hat{h}^{KS}(\mathbf{r})\phi_i(\mathbf{r}) = \varepsilon_i\phi_i(\mathbf{r}) \quad i = 1, 2, \ldots, n \tag{8.14}$$

where $\hat{h}^{KS}$ is the one-electron Kohn–Sham Hamiltonian (8.12).

With the best functionals available to date it is possible to obtain bond lengths within 0.01 Å for the diatomic molecules of the first-row atoms, and atomization energies within about $3\,\text{kcal}\,\text{mol}^{-1}$, at a cost which is sensibly lower than MP4 or other equivalent calculations.

9

Valence Bond Theory and the Chemical Bond

In this chapter we shall consider, first, elements of the Born–Oppenheimer approximation, concerning the separation in molecules of the motion of the electrons from that of the nuclei. It will be seen that, by neglecting small vibronic terms, the nuclei move in the field provided by the nuclei themselves and the molecular charge distribution of the electrons, determining what is called a potential energy surface, a function of the nuclear configuration.

Next, we shall introduce the study of the *chemical bond* by considering the simplest two-electron molecular example, the H_2 molecule. It will be seen that the single configuration MO approach fails to describe the correct dissociation of the molecule in ground-state H atoms because of the correlation error. A qualitatively correct description of the bond dissociation in H_2 is instead provided by the Heitler–London (HL) theory, where different electrons are allotted to different atomic orbitals, the resulting wavefunction for the ground state then being symmetrized with respect to electron interchange in order to satisfy Pauli's antisymmetry principle.

HL theory may be considered as introductory to the so-called *valence bond* (VB) theory of molecular electronic structure, where localized chemical bonds in molecules are described in terms of covalent and ionic structures. The theory is considered at an elementary level for giving qualitative help in studying the electronic structure of simple molecules, in a strict correspondence between quantum mechanical VB structures and chemical formulae. The importance of hybridization is stressed in

Methods of Molecular Quantum Mechanics: An Introduction to Electronic Molecular Structure
Valerio Magnasco
© 2009 John Wiley & Sons, Ltd

describing bond stereochemistry in polyatomic molecules, with particular emphasis on the H_2O molecule. A few applications of Pauling' semiempirical theory of π electrons in conjugated and aromatic hydrocarbons conclude the chapter.

9.1 THE BORN–OPPENHEIMER APPROXIMATION

This concerns the separation in molecules of the motion of the light electrons from the *slow* motion of the heavy nuclei.

We want to solve the molecular wave equation

$$\hat{H}\Psi = W\Psi \tag{9.1}$$

where $\hat{H}$ is the molecular Hamiltonian (in atomic units):

$$
\begin{aligned}
\hat{H} &= \sum_{\alpha} -\frac{1}{2M_\alpha}\nabla_\alpha^2 + \left(\sum_{i} -\frac{1}{2}\nabla_i^2 + V_{en} + V_{ee} \right) + V_{nn} \\
&= \sum_{\alpha} -\frac{1}{2M_\alpha}\nabla_\alpha^2 + \hat{H}_e + V_{nn}
\end{aligned}
\tag{9.2}
$$

In the expression above, the first term is the kinetic energy operator for the motion of the nuclei,[1] the term in parentheses is the electronic Hamiltonian $\hat{H}_e$ and the last term is the Coulombic repulsion between the point-like nuclei in the molecule.

Since wave equation (9.1) was too difficult to solve, Born and Oppenheimer (1927) suggested that, in a first approximation, the molecular wave function Ψ could be written as

$$\Psi(x, q) \approx \Psi_e(x; q)\Psi_n(q) \tag{9.3}$$

where Ψ_e is the electronic wavefunction, an ordinary function of the electronic coordinates x and *parametric* in the nuclear coordinates q. Ψ_e is a normalized solution of the electronic wave equation[2]

$$\hat{H}_e\Psi_e = E_e\Psi_e, \quad \langle \Psi_e | \Psi_e \rangle = 1 \tag{9.4}$$

---

[1] M_α is the mass of nucleus α in units of the electron mass.

[2] Which must be solved for any nuclear configuration specified by $\{q\}$.

Considering $\Psi_e\Psi_n$ as a nuclear variation function (Ψ_e = fixed), Longuet-Higgins (1961) showed that the *best* nuclear wavefunction Ψ_n satisfies the eigenvalue equation

$$\left[\sum_\alpha -\frac{1}{2M_\alpha}\nabla_\alpha^2 + \hat{U}_e(q)\right]\Psi_n(q) = W\Psi_n(q) \qquad (9.5)$$

where

$$\hat{U}_e(q) = E_e(q) + V_{nn} - \sum_\alpha \frac{1}{2M_\alpha}\int dx\ \Psi_e^*\nabla_\alpha^2\Psi_e$$
$$-\sum_\alpha \frac{1}{2M_\alpha}\int dx\ \Psi_e^*\nabla_\alpha\Psi_e \cdot \nabla_\alpha \qquad (9.6)$$

is the potential energy *operator* for the motion of the heavy nuclei in the electron cloud of the molecule. The assumption (9.3) about the molecular wavefunction is known as the first Born–Oppenheimer approximation. Consideration of just the first two terms in (9.6) gives what is known as the second Born–Oppenheimer approximation, where the last two small *vibronic terms* are omitted.[3] In this second approximation, the nuclear wave equation becomes:

$$\left[\sum_\alpha -\frac{1}{2M_\alpha}\nabla_\alpha^2 + U_e(q)\right]\Psi_n(q) = W\Psi_n(q) \qquad (9.7)$$

where $U_e(q)$ is now a purely multiplicative potential energy term. From (9.7) follows the possibility of defining a *potential energy surface* for the effective motion of the nuclei in the field provided by the nuclei themselves and the molecular electron charge distribution:

$$U_e(q) \approx E_e(q) + V_{nn} = E(q) \qquad (9.8)$$

Usually, we shall refer to (9.8) as the molecular energy in the Born–Oppenheimer approximation and refer to (9.7) as the Born–Oppenheimer nuclear wave equation, which determines the motion of the nuclei (e.g. in molecular vibrations) so familiar in spectroscopy.

---

[3] These terms can be neglected, in a first approximation, since they are of the order of $1/M_\alpha \approx 10^{-3}$.

The *adiabatic* approximation includes in $U_e(q)$ the third term in (9.6), which describes the effect of the nuclear Laplacian ∇^2_α on the electronic wavefunction Ψ_e. Both small vibronic terms in (9.6) can be included in a variational or perturbative way in a refined calculation of the molecular energy as a function of nuclear coordinates, and are responsible for interesting fine structural effects in the vibrational spectroscopy of polyatomic molecules (Jahn–Teller and Renner effects).

9.2   THE HYDROGEN MOLECULE H_2

Electrons and nuclei in the H_2 molecule are referred to the interatomic coordinate system of Figure 9.1. At A and B are the two protons (charge $+1$), a distance R apart measured along the z-axis; at 1 and 2 are the two electrons (charge -1). The bottom part of the figure shows the overlap

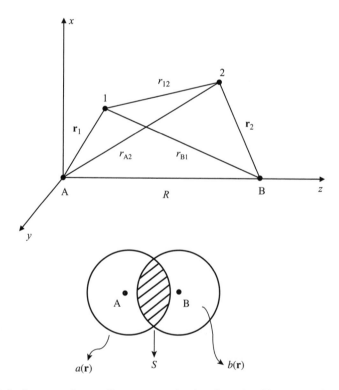

Figure 9.1   Interatomic coordinate system (top) and overlap S between spherical AOs (bottom)

S between two basic 1s STOs with orbital exponent c_0 (=1 for the free atoms):

$$a(\mathbf{r}_1) = \left(\frac{c_0^3}{\pi}\right)^{1/2} \exp(-c_0 r_1), \quad b(\mathbf{r}_2) = \left(\frac{c_0^3}{\pi}\right)^{1/2} \exp(-c_0 r_2) \qquad (9.9)$$

$$S = \langle b|a \rangle = (ab|1) = \exp(-\rho)\left(1 + \rho + \frac{1}{3}\rho^2\right) \quad \rho = c_0 R \qquad (9.10)$$

9.2.1 Molecular Orbital Theory

If σ_g and σ_u are the normalized bonding and antibonding one-electron MOs:

$$\sigma_g(\mathbf{r}) = \frac{a(\mathbf{r}) + b(\mathbf{r})}{\sqrt{2 + 2S}}, \quad \sigma_u = \frac{b(\mathbf{r}) - a(\mathbf{r})}{\sqrt{2 - 2S}} \qquad (9.11)$$

then the *one-configuration* two-electron MO wavefunction for the H$_2$ ground state is

$$\Psi(\text{MO}, {}^1\Sigma_g^+) = \|\sigma_g \bar{\sigma}_g\|$$
$$= \sigma_g(\mathbf{r}_1)\sigma_g(\mathbf{r}_2)\frac{1}{\sqrt{2}}[\alpha(s_1)\beta(s_2) - \beta(s_1)\alpha(s_2)] \quad S = M_S = 0$$

$$(9.12)$$

where the total spin quantum number S in (9.12) must not be confused with the overlap. In the molecular ground state, we allocate two electrons with opposite spin in the normalized spatial bonding MO $\sigma_g(\mathbf{r})$.

The MO energy in the Born–Oppenheimer approximation is

$$E(\text{MO}, {}^1\Sigma_g^+) = \langle \Psi(\text{MO}, {}^1\Sigma_g^+)|\hat{H}|\Psi(\text{MO}, {}^1\Sigma_g^+) \rangle$$
$$= \left\langle \sigma_g \sigma_g \middle| \hat{h}_1 + \hat{h}_2 + \frac{1}{r_{12}} + \frac{1}{R} \middle| \sigma_g \sigma_g \right\rangle = 2h_{\sigma_g \sigma_g} + (\sigma_g^2|\sigma_g^2) + \frac{1}{R}$$

$$(9.13)$$

where

$$\hat{h} = -\frac{1}{2}\nabla^2 - \frac{1}{r_A} - \frac{1}{r_B} \qquad (9.14)$$

is the bare-nuclei one-electron Hamiltonian.

In the hydrogenic approximation ($c_0 = 1$), we obtain for the matrix elements

$$
\begin{aligned}
2h_{\sigma_g\sigma_g} &= \frac{h_{aa} + h_{bb} + h_{ba} + h_{ab}}{1 + S} \\
&= 2E_H + \frac{(a^2|V_B) + (b^2|V_A) + (ab|V_B) + (ba|V_A)}{1 + S}
\end{aligned}
\qquad (9.15)
$$

$$(\sigma_g^2|\sigma_g^2) = \frac{\frac{1}{4}\left[(a^2|a^2) + (b^2|b^2)\right] + \frac{1}{2}(a^2|b^2) + (ab|ab) + (a^2|ba) + (b^2|ab)}{(1+S)^2} \qquad (9.16)$$

where both one- and two-electron integrals are written in charge density notation:

$$(a^2|V_B) = \langle a|-r_B^{-1}|a\rangle, \quad (ab|V_B) = \langle b|-r_B^{-1}|a\rangle \qquad (9.17)$$

$$(ab|ab) = \iint dr_1 dr_2 \frac{[a(r_2)b^*(r_2)]}{r_{12}}[a(r_1)b^*(r_1)] \qquad (9.18)$$

Because of the two-electron *monocentric* terms, $(a^2|a^2) = (b^2|b^2)$, the two-electron part (9.16) of the molecular energy is wrong when $R \to \infty$, since ($c_0 = 1$):

$$\lim_{R \to \infty}(\sigma_g^2|\sigma_g^2) = \frac{1}{2}(a^2|a^2) = \frac{1}{2} \times \frac{5}{8} = \frac{5}{16} = 0.3125E_h \qquad (9.19)$$

instead of zero, as it must be, since the interaction must vanish at infinite separation.

The interaction energy naturally decomposes into the following two terms:

$$\Delta E = E(MO, {}^1\Sigma_g^+) - 2E_H = \Delta E^{cb} + \Delta E^{exch-ov}({}^1\Sigma_g^+) \qquad (9.20)$$

$$\Delta E^{\mathrm{cb}} = (a^2|V_{\mathrm{B}}) + (b^2|V_{\mathrm{A}}) + (a^2|b^2) + \frac{1}{R} \qquad (9.21)$$

being the semiclassical Coulombic interaction energy and

$$\Delta E^{\mathrm{exch\text{-}ov}}(^1\Sigma_g^+) = \frac{(ab-Sa^2|V_{\mathrm{B}}) + (ba-Sb^2|V_{\mathrm{A}})}{1+S}$$

$$+ \frac{\frac{1}{4}[(a^2|a^2) + (b^2|b^2)] - \left(\frac{1}{2}+2S+S^2\right)(a^2|b^2) + (ab|ab) + (a^2|ab) + (b^2|ba)}{(1+S)^2}$$

$$(9.22)$$

the exchange-overlap component of the interaction energy, a quantum component depending on the electronic state of the molecule. The first term in (9.22) is the one-electron part, which has a correct behaviour when $R \to \infty$. In it appear the two exchange-overlap densities $a(\mathbf{r})b(\mathbf{r}) - Sa^2(\mathbf{r})$ and $b(\mathbf{r})a(\mathbf{r}) - Sb^2(\mathbf{r})$ (compare with Equation 4.17), having the property

$$\int d\mathbf{r}\,[a(\mathbf{r})b(\mathbf{r}) - Sa^2(\mathbf{r})] = 0, \quad \int d\mathbf{r}\,[b(\mathbf{r})a(\mathbf{r}) - Sb^2(\mathbf{r})] = 0 \qquad (9.23)$$

The second term in (9.22) is the two-electron part, which is incorrect when $R \to \infty$, due to the presence of the one-centre integrals. Therefore, we have for the ground state MO wavefunction

$$\lim_{R \to \infty} \Delta E(\mathrm{MO}, {}^1\Sigma_g^+) = \frac{1}{2}(a^2|a^2) \qquad (9.24)$$

corresponding to the erroneous dissociation

$$\mathrm{H}_2({}^1\Sigma_g^+) \to \mathrm{H}(^2\mathrm{S}) + \frac{1}{2}\mathrm{H}^-({}^1\mathrm{S}) \qquad (9.25)$$

instead of the correct one:

$$\mathrm{H}_2({}^1\Sigma_g^+) \to 2\mathrm{H}(^2\mathrm{S}) \qquad (9.26)$$

The behaviour of the potential energy curve for the MO description of ground-state H$_2$ is qualitatively depicted in the top part of Figure 9.2, which shows the asymptotically incorrect behaviour of the dissociation

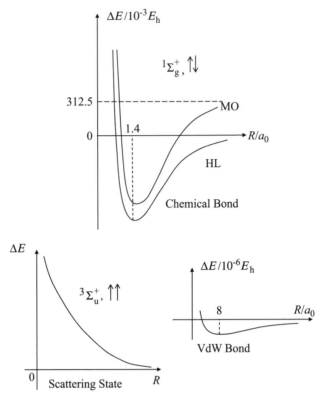

Figure 9.2   MO and HL potential energy curves for $^1\Sigma_g^+$ ground state (top) and $^3\Sigma_u^+$ excited state (bottom) of the H_2 molecule

energy. This large *correlation error* can be removed through CI of the ground-state σ_g^2 configuration with the doubly excited configuration σ_u^2, where both electrons occupy the σ_u antibonding MO, leaving σ_g empty. The *interconfigurational* wavefunction

$$\Psi(\text{MO-CI}, {}^1\Sigma_g^+) = N[\Psi(\sigma_g^2, {}^1\Sigma_g^+) + \lambda\Psi(\sigma_u^2, {}^1\Sigma_g^+)] \qquad (9.27)$$

with the variationally optimized mixing parameter $\lambda \approx -0.13$ at $R_e = 1.4a_0$, now correctly describes dissociation of the molecule into two neutral H atoms in their ground state.

9.2.2   Heitler–London Theory

Heitler and London (1927) proposed for H_2 the *two-configuration* wavefunction

$$\Psi(\mathrm{HL}, {}^1\Sigma_g^+) = N\{\|a\bar{b}\| - \|\bar{a}b\|\}$$

$$= \frac{a(\mathbf{r}_1)b(\mathbf{r}_2) + b(\mathbf{r}_1)a(\mathbf{r}_2)}{\sqrt{2 + 2S^2}} \frac{1}{\sqrt{2}}[\alpha(s_1)\beta(s_2) - \beta(s_1)\alpha(s_2)]$$

$$S = M_S = 0 \tag{9.28}$$

giving the HL energy in the Born–Oppenheimer approximation:

$$
\begin{aligned}
E(\mathrm{HL}, {}^1\Sigma_g^+) &= \langle \Psi(\mathrm{HL}, {}^1\Sigma_g^+) | \hat{H} | \Psi(\mathrm{HL}, {}^1\Sigma_g^+) \rangle \\
&= \left\langle \frac{ab+ba}{\sqrt{2+2S^2}} \left| \hat{h}_1 + \hat{h}_2 + \frac{1}{r_{12}} + \frac{1}{R} \right| \frac{ab+ba}{\sqrt{2+2S^2}} \right\rangle \\
&= \frac{h_{aa} + h_{bb} + S(h_{ba} + h_{ab})}{1+S^2} + \frac{(a^2|b^2) + (ab|ab)}{1+S^2} + \frac{1}{R}
\end{aligned}
$$

$$\tag{9.29}$$

It is seen that the HL two-electron component of the molecular energy is much simpler than its MO counterpart and now has the correct behaviour as $R \to \infty$.

In the hydrogenic approximation ($c_0 = 1$), the HL interaction energy for ${}^1\Sigma_g^+$ H₂ is

$$\Delta E(\mathrm{HL}, {}^1\Sigma_g^+) = E(\mathrm{HL}, {}^1\Sigma_g^+) - 2E_\mathrm{H} = \Delta E^{\mathrm{cb}} + \Delta E^{\mathrm{exch\text{-}ov}}({}^1\Sigma_g^+) \tag{9.30}$$

where ΔE^{cb} is the *same* as the MO expression (9.21), while

$$\Delta E^{\mathrm{exch\text{-}ov}}({}^1\Sigma_g^+) = S\frac{(ab - Sa^2|V_\mathrm{B}) + (ba - Sb^2|V_\mathrm{A})}{1+S^2} + \frac{(ab|ab) - S^2(a^2|b^2)}{1+S^2}$$

$$\tag{9.31}$$

Both components of the interaction energy now vanish as $R \to \infty$, therefore describing correctly the dissociation of the ground-state H₂ molecule into neutral ground-state H atoms (top part of Figure 9.2, HL curve).

For the excited state of the H₂ molecule, we have the *triplet* wavefunction

$$\Psi(\mathrm{HL}, {}^3\Sigma_u^+) = \begin{cases} \|a\, b\| & S=1,\ M_S = 1 \\[2mm] \dfrac{1}{\sqrt{2}}[\|a\,\bar{b}\| + \|\bar{a}\, b\|] & S=1,\ M_S = 0 \\[2mm] \|\bar{a}\,\bar{b}\| & S=1,\ M_S = -1 \end{cases} \tag{9.32}$$

or, by expanding the determinants:

$$\Psi(\mathrm{HL}, ^3\Sigma_\mathrm{u}^+) = \frac{a(\mathbf{r}_1)b(\mathbf{r}_2) - b(\mathbf{r}_1)a(\mathbf{r}_2)}{\sqrt{2 - 2S^2}} \begin{cases} \alpha(s_1)\alpha(s_2) \\ \dfrac{1}{\sqrt{2}}[\alpha(s_1)\beta(s_2) + \beta(s_1)\alpha(s_2)] \\ \beta(s_1)\beta(s_2) \end{cases}$$

$$(9.32')$$

The quantum mechanical exchange-overlap component of the interaction energy for the triplet is now

$$\Delta E^{\text{exch-ov}}(^3\Sigma_\mathrm{u}^+) = -S \frac{(ab - Sa^2|V_\mathrm{B}) + (ba - Sb^2|V_\mathrm{A})}{1 - S^2} - \frac{(ab|ab) - S^2(a^2|b^2)}{1 - S^2}$$

$$(9.33)$$

and is repulsive for any value of R (a scattering state, left bottom part of Figure 9.2). At the rather large internuclear distance of about $8a_0$ we observe the formation of a rather flat van der Waals (VdW) minimum of $-20 \times 10^{-6} E_\mathrm{h}$, due to the prevalence of the weak London attractive forces (Chapter 11) over this weak triplet repulsion. The appearance of this weak VdW bond is depicted qualitatively in the right bottom part of Figure 9.2, which refers to large internuclear distances.[4] It should be noted that, at variance with the ground state, $\Psi(\mathrm{MO}, ^3\Sigma_\mathrm{u}^+)$ now coincides with $\Psi(\mathrm{HL}, ^3\Sigma_\mathrm{u}^+)$.

9.3   THE ORIGIN OF THE CHEMICAL BOND

The HL theory of the chemical bond in H_2 can be considered as the first approximation including exchange in an RS perturbation expansion of the interatomic energy (see Chapters 10 and 11). In a VB calculation, variational optimization of the orbital exponent c_0 of a simple STO basis is tantamount to including in first order a large part of the spherical distortion, so giving a fairly good description of the quantum exchange-overlap component of the interaction that is dominant in the bond region (Magnasco, 2007).

---

[4] Note in the figure the change in the energy scale for chemical (top, $10^{-3}E_\mathrm{h}$) and VdW (bottom, $10^{-6}E_\mathrm{h}$) bonds, meaning that the latter are 1000 times smaller in their order of magnitude.

Table 9.1 ΔE bond energies $(10^{-3}E_{\mathrm{h}})$ for the ground state of simple homonuclear diatomics

R/a_0	$H_2^+\ (^2\Sigma_g^+)$	$H_2\ (^1\Sigma_g^+)$	$He_2^+\ (^2\Sigma_u^+)$
1	59.001 6	−88.585 2	570.468 1
1.2	−15.921 6	−128.956 1	223.806 2
1.4	−55.339 8	−139.049 5	46.991 8
1.6	−75.319 2	−134.372 6	−39.864 2
1.8	−84.176 4	−122.673 4	−78.053 4
2	−86.506 0	−108.010 1	−89.992 0
3	−64.448 2	−41.699 3	−46.940 5
4	−37.334 5	−11.358 2	−13.819 3
5	−19.205 4	−2.534 91	−3.421 80
6	−9.080 4	−0.509 89	−0.783 36

To investigate the origin of the chemical bond, optimized VB calculations were done (Magnasco, 2008) for the ground states of $H_2^+\ (^2\Sigma_g^+)$, $H_2\ (^1\Sigma_g^+)$ and $He_2^+\ (^2\Sigma_u^+)$, a series of simple molecules that are the prototypes of one-, two- and three-electron σ chemical bonds, and of the Pauli repulsion for the He(1s)–He(1s) interaction. The bond energy ΔE is analysed into its polarized Coulombic ΔE^{cb} and exchange-overlap $\Delta E^{\mathrm{exch\text{-}ov}}$ components, the Coulombic component being best obtained by direct analytical calculation avoiding the round-off errors that would arise at the larger distances.

A summary of the results[5] is presented in Tables 9.1–9.4. For all systems studied, including the molecule-ions H_2^+ and He_2^+, the Coulombic component (Table 9.2) has a charge-overlap nature rapidly decreasing

Table 9.2 ΔE^{cb} Coulombic component $(10^{-3}E_{\mathrm{h}})$ of the bond energies for the ground state of simple homonuclear diatomics

R/a_0	H_2^+	H_2	He_2^+
1	117.092 9	15.846 9	41.179 9
1.2	68.997 0	−9.928 7	−5.603 9
1.4	42.746 1	−19.423 1	−14.892 8
1.6	27.486 5	−21.831 1	−13.535 4
1.8	18.173 0	−21.076 3	−9.975 0
2	12.260 2	−18.993 9	−6.709 6
3	2.005 7	−7.016 22	−0.522 67
4	0.341 95	−1.677 58	−0.026 28
5	0.053 3	−0.329 08	−0.001 10
6	0.007 5	−0.060 06	−0.000 04

---

[5] The optimized values of the orbital exponent c_0 at the various R are given in Magnasco (2008).

Table 9.3 $\Delta E^{\text{exch-ov}}$ exchange-overlap component ($10^{-3}E_{\text{h}}$) of the bond energies for the ground state of simple homonuclear diatomics

R/a_0	H_2^+ ($^2\Sigma_{\text{g}}^+$)	H_2($^1\Sigma_{\text{g}}^+$)	He_2^+ ($^2\Sigma_{\text{u}}^+$)
1	−58.091 3	−104.432 1	529.288 2
1.2	−84.918 6	−119.027 4	229.410 1
1.4	−98.085 9	−119.626 4	61.884 6
1.6	−102.805 7	−112.541 5	−26.328 8
1.8	−102.349 4	−101.597 1	−68.078 4
2	−98.766 2	−89 016 2	−83.282 4
3	−66.453 9	−34.683 1	−46.417 8
4	−37.676 4	−9.680 6	−13.793 0
5	−19.258 7	−2.205 8	−3.420 7
6	−9.087 9	−0.449 83	−0.783 32

with the internuclear distance R as $\exp(-c_0 R)$ and being always repulsive for H_2^+. This means that, classically, no bond can be formed between a ground-state H atom and a proton H^+. For the remaining systems, ΔE^{cb} is repulsive at small values and attractive at large values of R, but always insufficient to originate a sufficiently stable chemical bond. For all systems, the greater part of the bond energy comes from the exchange-overlap contribution between spherically polarized charge distributions (Tables 9.3 and 9.4), which, in the bond region, is attractive for H_2^+, H_2 and He_2^+ and repulsive for the He–He interaction.

Hence, we conclude that, for all chemically bonded molecules, the dominant component of the interaction in the bond region is the *exchange-overlap* component, which depends on the spin coupling of the interacting atoms; this is always *attractive* when forming the chemical bond (Table 9.3) and is *repulsive* when bonding is forbidden (Table 9.4). So, the exchange-overlap component is at the origin of both the chemical bond and Pauli repulsion. In an appendix given as an electronic addition

Table 9.4 ΔE Pauli repulsion and its Coulombic and exchange-overlap components ($10^{-3}E_{\text{h}}$) for the He(1s)–He(1s) interaction in the medium range of interatomic separations

R/a_0	ΔE	ΔE^{cb}	$\Delta E^{\text{exch-ov}}$
2	136.616	−27.282 1	163.898
2.5	42.669 3	−7.549 17	50.218 5
3	12.964 0	−1.930 70	14.894 7
3.5	3.799 79	−0.468 75	4.268 54
4	1.073 17	−0.109 52	1.182 69
4.5	0.292 81	−0 024 864	0.317 67
5	0.077 48	−0.005 522	0.083 00

to the paper quoted above (Magnasco, 2008), the origin of the quantum mechanical exchange-overlap densities was studied in detail, showing their different behaviour in the case of the chemical bond in ground-state H_2 and of the Pauli repulsion between two ground-state He atoms. This made clear the relation between such VB 'exchange-overlap' densities and the MO density 'interference' terms introduced time ago by Ruedenberg (1962).

The values at the minima of the potential energy calculations, not shown in the tables, are $-86.506 \times 10^{-3} E_h$ at $R_e = 2a_0$ with $c_0 = 1.238$ for H_2^+ ($^2\Sigma_g^+$), $-139.079 \times 10^{-3} E_h$ at $R_e = 1.41a_0$ with $c_0 = 1.165$ for H_2 ($^1\Sigma_g^+$), $-90.498 \times 10^{-3} E_h$ at $R_e = 2.06a_0$ with $c_0 = 1.832$ for He_2^+ ($^2\Sigma_u^+$), while the He(1s)–He(1s) interaction is repulsive by $12.964 \times 10^{-3} E_h$ at $R = 3a_0$ with $c_0 = 1.691$. In this way, it is seen that practically correct R_e bond lengths and about 85% of the experimentally observed or accurately calculated ΔE_e bond energies (Peek, 1965; Huber and Herzberg, 1979 for H_2^+; Wolniewicz, 1993 for H_2; Cencek and Rychlewski, 1995 for He_2^+) are obtained from these optimized calculations, while the calculated Pauli repulsion for He(1s)–He(1s) in the medium range at, say, $R = 3a_0$, gives over 95% of the correct repulsion (Liu and McLean, 1973). It should be remarked that the optimized orbital exponents c_0 *contract* over the hydrogenic value $Z = 1$ for H_2^+ and H_2 and *expand* over $Z = 2$ for He_2^+ and He_2. This change in shapes of the optimized AOs suggests that the guiding principle in the elementary VB theory of the chemical bond is the maximum of the exchange-overlap energy and *not* the maximum overlap, as discussed elsewhere (Magnasco and Costa, 2005).

9.4   VALENCE BOND THEORY AND THE CHEMICAL BOND

This is a *multi-determinant* theory which originates from the HL theory of the H_2 molecule just discussed above.

9.4.1   Schematization of Valence Bond Theory

Basis of (spatial) AOs $\Rightarrow$ atomic spin-orbitals (ASOs) $\Rightarrow$ antisymmetrized products (APs), that is, Slater determinants of order equal to the number N of the electrons in the molecule, eigenfunctions of the $\hat{S}_z$ operator with eigenvalue $M_S = (N_\alpha - N_\beta)/2 \Rightarrow$ VB structures, eigenstates of $\hat{S}^2$ with eigenvalue $S(S + 1)$ (defined S) $\Rightarrow$ symmetry combinations of VB structures $\Rightarrow$ multi-determinant wavefunctions describing at the full-VB level the electronic states of the molecule.

9.4.2 Schematization of Molecular Orbital Theory

Basis of (spatial) AOs $\Rightarrow$ MOs by the LCAO method, classified according to symmetry-adapted types (Chapter 12) $\Rightarrow$ single Slater determinant of doubly occupied MOs (for closed shells) $\Rightarrow$ MO-CI among all determinants belonging to the same symmetry $\Rightarrow$ multi-determinant wavefunctions describing at the MO-full-CI level the electronic states of the molecule.

Starting from the *same* AO basis, full-VB and MO-full-CI methods are *entirely equivalent* at the end of each process, but may be deeply different in their early stages (as we have seen for the H_2 molecule).

9.4.3 Advantages of the Valence Bond Method

1. VB structures (such as H–H, H^-H^+, H^+H^-, Li–H, Li^+-H^-, $N \equiv N$, $C^- \equiv O^+$) are related to the existence of *chemical bonds* in molecules, the most important ones corresponding to the rule of the so-called 'perfect-pairing of bonds' (see Figure 9.6 for H_2O).
2. Principle of maximum overlap (better, minimum of exchange-overlap bond energy) $\Rightarrow$ bond stereochemistry in polyatomic molecules (Magnasco and Costa, 2005).
3. It allows for a correct description of the dissociation of the chemical bonds (of crucial importance for studying chemical reactions).
4. It gives a sufficiently accurate description of the spin densities in radicals (see the example of the allyl radical in Section 9.6).
5. For small molecules, the ab initio formulation of the theory allows one to account up to about 80% of the electronic correlation and to get bond distances within $0.02 a_0$.

9.4.4 Disadvantages of the Valence Bond Method

1. Nonorthogonal basic AOs $\Rightarrow$ nonorthogonal VB structures $\Rightarrow$ difficulties in the evaluation of the matrix elements of the Hamiltonian.

Table 9.5 Comparison between the degree of Hückel and covalent VB secular equations for the singlet state of some polycyclic hydrocarbons

Molecule	N	$n = N/2$	f_0^N
Benzene	6	3	5
Naphthalene	10	5	42
Anthracene	14	7	429
Coronene	24	12	208 012

2. The number of covalent VB structures of given total spin S increases rapidly with the number $n = N/2$ of bonds, according to the Wigner's (1959) formula:

$$f_S^N = \binom{2n}{n-S} - \binom{2n}{n-S-1} = \frac{(2S+1)(2n)!}{(n+S+1)!(n-S)!} \qquad (9.34)$$

Table 9.5 gives a comparison between the order of Hückel and covalent VB secular equations for the singlet state of some polycyclic hydrocarbons whose structure is reported in Figure 9.3. When N is increasing, there is a striking difference between the second and the last column of the table, giving the order of the respective secular equations.

The situation is even worse if we take into consideration ionic structures as well. In this case, the total number of structures, covalent plus all possible ionic, is given by Weyl's formula (Mulder, 1966):

$$f(N, m, S) = \frac{2S+1}{m+1} \binom{m+1}{\frac{N}{2}+S+1} \binom{m+1}{\frac{N}{2}-S} \qquad (9.35)$$

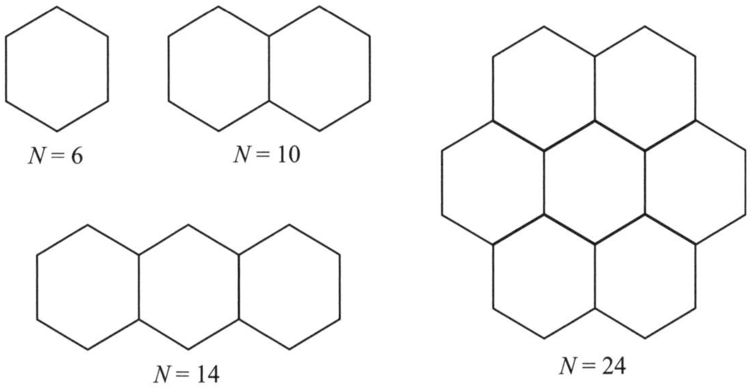

$N = 6$ $N = 10$

$N = 14$ $N = 24$

Figure 9.3 Structure of some polycyclic hydrocarbons. From top left: benzene, naphthalene, anthracene, coronene

where N is the number of electrons, m the number of basic AOs and S the total spin. Recall the formula for the binomial coefficient:

$$\binom{n}{m} = \frac{n!}{m!(n-m)!} \quad n \geq m \tag{9.36}$$

Consider the example of the π-electron system of benzene:

(i) Single zeta (SZ) basis set

$$N = 6, \; m = 6, \; S = 0 \Rightarrow f(6,6,0) = 175$$

Singlet VB structures (5 covalent + 170 ionic).

(ii) Double zeta (DZ) basis set

$$N = 6, \; m = 12, \; S = 0 \Rightarrow f(6,12,0) = 15\,730$$

The total number of possible VB structures increases very rapidly with the size of the basic AOs!

9.4.5   Construction of Valence Bond Structures

VB structures are specified by giving their *parent*, a Slater determinant where the spin-orbitals involved in the bond must have opposite spin. In brief notation, we shall represent the parent by indicating in parentheses only those valence orbitals that are bonded together, omitting all core orbitals which, in a first approximation, are considered as not taking part in the bond. The VB structures (eigenstates of $\hat{S}^2$ belonging to the total spin S) are then simply obtained from the parents by doing all possible spin interchanges between the bonded orbitals, with a minus sign introduced for each interchange.

(i) LiH ($^1\Sigma^+$)

$$\text{Li}(^2\text{S}) : 1s^2_{\text{Li}}|s \quad \text{H}(^2\text{S}) : \text{h}$$

Covalent structure	Ionic structures	
Li–H	Li$^+$H$^-$	Li$^-$H$^+$
(s$\bar{\text{h}}$)	(h$\bar{\text{h}}$)	(s$\bar{\text{s}}$)

where the parents are[6]

$$(s\bar{h}) = ||1s_{Li}1\bar{s}_{Li}s\bar{h}||, \quad (h\bar{h}) = ||1s_{Li}1\bar{s}_{Li}h\bar{h}||, \quad (s\bar{s}) = ||1s_{Li}1\bar{s}_{Li}s\bar{s}|| \tag{9.37}$$

the first denoting a covalent bond between Li and H, the second the ionic structure Li^+H^- (both electrons on H) and the last the ionic structure Li^-H^+ (both electrons on Li). The full-VB structure denoting covalent–ionic resonance in LiH will be

$$\begin{aligned}\Psi(\text{LiH},{}^1\Sigma^+) &= \psi_1 c_1 + \psi_2 c_2 + \psi_3 c_3 \\ &\propto [(s\bar{h})-(\bar{s}h)] + \lambda_1(h\bar{h}) + \lambda_2(s\bar{s})\end{aligned} \tag{9.38}$$

where ψ_1 is the (normalized) singlet covalent structure associated with the parent $(s\bar{h})$:

$$\psi_1 = \frac{1}{\sqrt{2}}\left[(s\bar{h})-(\bar{s}h)\right] \tag{9.39}$$

while the ionic parents $(h\bar{h})$ and $(s\bar{s})$ are already eigenstates of $\hat{S}^2$ with eigenvalue $S=0$:

$$\psi_2 = (h\bar{h}), \quad \psi_3 = (s\bar{s}) \tag{9.40}$$

The mixing coefficients in (9.38) are determined by the Ritz method by solving the appropriate (3×3) secular equation. Because of the strong difference in electronegativity between Li and H, it is reasonable to expect that $\lambda_1 \gg \lambda_2$.

(ii) FH $({}^1\Sigma^+)$

$$F({}^2P) : 1s_F^2\, 2s_F^2\, 2p\pi_F^4 | 2p\sigma_F \quad H({}^2S) : h$$

Covalent structure	Ionic structures	
F–H	F^-H^+	F^+H^-
$(\sigma_F\bar{h})$	$(\sigma_F\bar{\sigma}_F)$	$(h\bar{h})$

---

[6] Parent determinants have all properties of complete Slater determinants.

where now ($N = 10$) the first parent stands for the (10×10) Slater determinant:

$$(\sigma_F \bar{h}) = ||1s_F 1\bar{s}_F 2s_F 2\bar{s}_F 2p\pi_{xF} 2p\bar{\pi}_{xF} 2p\pi_{yF} 2p\bar{\pi}_{yF}|2p\sigma_F \bar{h}|| \quad (9.41)$$

the singlet covalent VB structure being

$$\psi_1 = \frac{1}{\sqrt{2}} [(\sigma_F \bar{h}) - (\bar{\sigma}_F h)] \quad (9.42)$$

and so on.

(iii) $N_2(^1\Sigma_g^+)$

$$N_A(^4S) : 1s_{N_A}^2 2s_{N_A}^2 |\sigma_A x_A y_A \quad \sigma = 2p\sigma = 2p_z \quad N = 7$$

$$N_B(^4S) : 1s_{N_B}^2 2s_{N_B}^2 |\sigma_B x_B y_B \quad x = 2p\pi_x = 2p_x \quad N = 7$$

If, in a first approximation, we neglect sp hybridization onto the N atoms, then we can describe as follows the formation of the three N–N bonds in N_2 (top row of Figure 9.4).

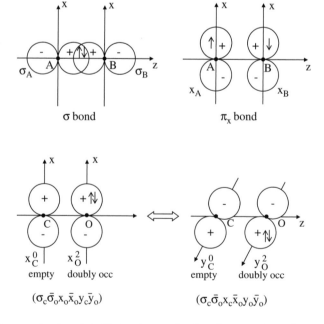

Figure 9.4 σ and π_x bonds in N_2 (top), π_x and π_y bonds in CO from charge transfer O to C (bottom)

Covalent σ bond (directed along the internuclear z-axis):

$$(\sigma_A \bar{\sigma}_B) - (\bar{\sigma}_A \sigma_B)$$

Covalent π_x bond (perpendicular to the internuclear axis, directed along x):

$$(x_A \bar{x}_B) - (\bar{x}_A x_B)$$

Covalent π_y bond (perpendicular to the internuclear axis, directed along y):

$$(y_A \bar{y}_B) - (\bar{y}_A y_B)$$

Altogether these describe the *covalent triple bond* in N_2 in terms of the parent:

$$(\sigma_A \bar{\sigma}_B x_A \bar{x}_B y_A \bar{y}_B) \tag{9.43}$$

The full singlet $(S = M_S = 0)$ covalent VB structure of N_2 is then given by the combination of the eight Slater determinants:

$$
\begin{aligned}
\Psi(^1\Sigma_g^+) = (\sqrt{8})^{-1}[&(\sigma_A \bar{\sigma}_B x_A \bar{x}_B y_A \bar{y}_B) - (\bar{\sigma}_A \sigma_B x_A \bar{x}_B y_A \bar{y}_B) \\
& - (\sigma_A \bar{\sigma}_B \bar{x}_A x_B y_A \bar{y}_B) - (\sigma_A \bar{\sigma}_B x_A \bar{x}_B \bar{y}_A y_B) \\
& + (\bar{\sigma}_A \sigma_B \bar{x}_A x_B y_A \bar{y}_B) + (\bar{\sigma}_A \sigma_B x_A \bar{x}_B \bar{y}_A y_B) \\
& + (\sigma_A \bar{\sigma}_B \bar{x}_A x_B \bar{y}_A y_B) - (\bar{\sigma}_A \sigma_B \bar{x}_A x_B \bar{y}_A y_B)]
\end{aligned}
\tag{9.44}
$$

where orthogonality is assumed between the determinants, the first term (I) being a shorthand for the (normalized) complete (14×14) Slater determinant:

$$(I) = ||1s_A 1\bar{s}_A 1s_B 1\bar{s}_B \, 2s_A 2\bar{s}_A 2s_B 2\bar{s}_B | \sigma_A \bar{\sigma}_B x_A \bar{x}_B y_A \bar{y}_B|| \tag{9.45}$$

In (9.45), the first part on the left describes eight electrons in frozen-core AOs and the last part the six valence electrons engaged in the triple bond.

(iv) CO $(^1\Sigma^+)$

$$C(^3P) : 1s_C^2 2s_C^2 | \sigma_C x_C y_C^0 \quad \text{or} \quad x_C^0 y_C \quad N = 6$$

$$O(^3P) : 1s_O^2 2s_O^2 | \sigma_O x_O y_O^2 \quad \text{or} \quad x_O^2 y_O \quad N = 8$$

The CO molecule is a heteropolar molecule isoelectronic with N_2. Since electron charge transfer can occur between the *doubly occupied* AOs on oxygen towards the corresponding *empty* AOs on carbon (bottom row of Figure 9.4), the most probable structure of CO is $C^- \equiv O^+$, which describes formation of an *ionic triple bond* as shown by the parent:

$$(\sigma_C \bar{\sigma}_O x_C \bar{x}_O y_C \bar{y}_O) \qquad (9.46)$$

(v) He_2^+ ($^2\Sigma_u^+$) three-electron σ bond

While He_2 is a VdW molecule (see Chapter 11), He_2^+ is a rather stable diatomic molecule with $D_e = 90.78 \times 10^{-3} E_h$ at $R_e = 2.043 a_0$ (Huber and Herzberg, 1979), which can be considered the prototype of the three-electron σ bond. With reference to the top row of Figure 9.5, the single normalized doublet VB structures $(S = M_S = \frac{1}{2})$ are

$$\psi_1 = N_1[(a\bar{a}b) - (aa\bar{b})] = ||a\bar{a}b|| \quad N_1 = [6(1-S^2)]^{-1/2} \quad (9.47)$$

$$\psi_2 = N_2[(b\bar{b}a) - (bb\bar{a})] = ||b\bar{b}a|| \quad N_2 = [6(1-S^2)]^{-1/2} \quad (9.48)$$

where both the second determinants in the square brackets of either (9.47) or (9.48) are zero because of Pauli's principle (two spin-orbitals are equal).

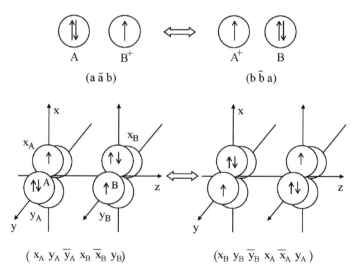

Figure 9.5 Three-electron σ bond in He_2^+ (top) and three-electron π bonds in ground-state O_2 (bottom)

The normalized full-VB doublet wavefunction with the correct symmetry for the $^2\Sigma_u^+$ ground state is

$$\Psi(^2\Sigma_u^+) = \frac{\psi_1 - \psi_2}{\sqrt{2 + 2S}} \qquad (9.49)$$

while that for the excited $^2\Sigma_g^+$ state[7] is

$$\Psi(^2\Sigma_g^+) = \frac{\psi_1 + \psi_2}{\sqrt{2 - 2S}} \qquad (9.50)$$

Using elementary rules of determinants, it can be easily shown that, in this case, the VB and MO wavefunctions coincide:

$$\Psi(MO, {}^2\Sigma_u^+) = \|\sigma_g \,\bar{\sigma}_g \,\sigma_u\| = \Psi(VB, {}^2\Sigma_u^+) \qquad (9.51)$$

$$\Psi(MO, {}^2\Sigma_g^+) = \|\sigma_g \,\sigma_u \,\bar{\sigma}_u\| = \Psi(VB, {}^2\Sigma_g^+) \qquad (9.52)$$

Recent ab initio calculation (Magnasco, 2008) of the potential energy curve for ground-state He_2^+ with a minimal basis set of optimized 1s STOs gives $D_e = 90.49 \times 10^{-3} E_h$ at $R_e = 2.05 a_0$ for $c_0 = 1.833$, in excellent agreement with the spectroscopically observed values given above. The excited $^2\Sigma_g^+$ state is always repulsive in the bond region, giving a scattering state.

(vi) $O_2\,(^3\Sigma_g^-)$ three-electron π bonds

$$O_A(^3P) : 1s_A^2 2s_A^2 |\sigma_A x_A y_A^2 \quad \text{or} \quad x_A^2 y_A \quad N = 8$$

$$O_B(^3P) : 1s_B^2 2s_B^2 |\sigma_B x_B y_B^2 \quad \text{or} \quad x_B^2 y_B \quad N = 8$$

The electronic structure of this 16-electron paramagnetic molecule depends entirely on the system of its π electrons (six electrons in four $2p\pi$ orbitals), while there is always one σ bond between A and B directed along the internuclear z-axis. The two possible *singlet* VB structures for the ground state would imply formation of a π bond and two parallel $2p\pi$ lone pairs containing four electrons, having the parents

$$(x_A \bar{x}_B \, y_A \bar{y}_A \, y_B \bar{y}_B) \Leftrightarrow (x_A \bar{x}_A \, x_B \bar{x}_B \, y_A \bar{y}_B) \qquad (9.53)$$

---

[7] The symmetries of ground and excited state in He_2^+ are exactly opposed to those observed for H_2^+.

The corresponding structures are highly improbable, however, because of the strong Pauli repulsion existing between coplanar lone pairs. The most probable VB structures are obtained by twisting by 90° the two lone pairs (bottom row of Figure 9.5) so that Pauli repulsion is drastically reduced, leading to the parents

$$(x_A y_A \, \bar{y}_A x_B \, \bar{x}_B y_B) \Leftrightarrow (x_B y_B \, \bar{y}_B x_A \, \bar{x}_A y_A) \qquad (9.54)$$

Full resonance between the corresponding two equivalent structures describes now the formation of two highly stable three-electron π bonds (Wheland, 1937), yielding for the ground state of O_2 a *triplet* $^3\Sigma_g^-$.

9.5   HYBRIDIZATION AND MOLECULAR STRUCTURE

9.5.1   The H_2O Molecule

Let us consider now the formation of two equivalent O—H bonds in the H_2O molecule making an interbond angle 2θ of about 105°. Hybridization, namely the mixing of sp AOs onto the *same* nucleus, is now essential in order to give straight bonds satisfying the principle of maximum overlap. With reference to the top part of Figure 9.6, we see that the two 2p AOs on oxygen directed along the bonds are no longer orthogonal to each other:

$$\langle p_1 | p_2 \rangle = \cos 2\theta = -0.25882 \qquad (9.55)$$

It is evident from (9.55) that we can have orthogonality only for $2\theta = 90°$. We can restore orthogonality of directed orbitals on oxygen simply by mixing in some s AO, and a detailed calculation shows that the appropriately directed sp^2 (C_{2v}) hybrids on O are (top part of Figure 9.6)

$$\begin{cases} hy_1 = 0.4534s + 0.5426z + 0.7071y = b_1 \\ hy_2 = 0.4534s + 0.5426z - 0.7071y = b_2 \\ hy_3 = 0.7673s - 0.6412z = \sigma \end{cases} \qquad (9.56)$$

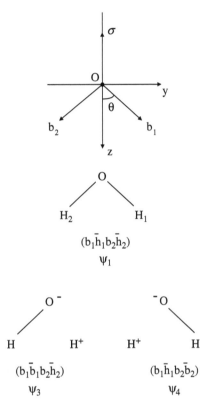

Figure 9.6 The three sp^2 hybrids of C_{2v} symmetry (top) and the most important covalent and singly polar VB structures in H_2O (bottom)

giving 20.6% s and 79.4% p for the hybrids engaged in the two equivalent O−H bonds and 58.9% s and 41.1% p for the hybrid directed away on the back of the z-axis and forming the σ lone pair.

We can now consider the formation of covalent and ionic structures involving such directed hybrids, limiting ourselves to consideration of the four-electron problem, leaving inner shell and lone pair electrons *frozen* (only a first approximation). In an obvious brief notation:

$$O(^3P): 1s_O^2 2s_O^2 x_O^2 y_O z_O \Rightarrow k^2 \sigma^2 x^2 b_1 b_2 \quad H(^2S): h_1, h_2 \quad (9.57)$$

Weyl's formula ($N = 4$, $m = 4$, $S = 0$) gives in this case

$$f(4, 4, 0) = \frac{1}{5}\binom{5}{3}\binom{5}{2} = 20 \quad (9.58)$$

so that there are altogether 20 singlet VB structures (2 covalent, 12 singly polar, 6 doubly polar), the most important being those given in the bottom part of Figure 9.6. An approximate VB function describing correctly the observed value of the dipole moment ($\mu = 0.73ea_0$) was estimated by Coulson (1961) as:

$$\Psi(^1A_1) = \psi_1 c_1 + \frac{1}{\sqrt{2}}(\psi_3 + \psi_4)c_3 + \psi_{15}c_{15} \qquad (9.59)$$

with $c_1 = 0.64$, $c_3 = c_4 = 0.48$ and $c_{15} = 0.36$, where ψ_{15} is the doubly polar structure with all four electrons on O ($b_1\bar{b}_1 b_2 \bar{b}_2$). Wavefunction (9.59) yields the following relative percentage weights for the VB structures:

$$41\,\%\,\psi_1, \quad 23\,\%\,\psi_3, \quad 23\,\%\,\psi_4, \quad 13\,\%\,\psi_{15} \qquad (9.60)$$

9.5.2 Properties of Hybridization

- Hybrid AOs (see b_A and b_B in the bottom part of Figure 9.7) become *unsymmetrical* with respect to their centres, acquiring what Coulson (1961) calls an atomic dipole.[8]
- Physically, hybridization describes *polarization* (distortion) of the AOs engaged in the bond. As such, it may involve mixing of occupied and empty AOs.[9]
- Hybridization restores orthogonality of AOs on the same atom, allowing for interhybrid angles *greater* than 90°. Hybrid AOs can in this way reorient themselves in an optimum way, yielding orbitals directed along the bonds and avoiding formation of weaker bent bonds.
- Hybridization gives in this way the AOs the appropriate directional character for forming strong covalent bonds and, therefore, is of fundamental importance in stereochemistry.
- Even without changes in valency (*isovalent* hybridization), hybridization allows for a better disposition in space of electron lone pairs, projecting them outside the bond region (as in the H_2O molecule), which has a considerable effect on the electric dipole moment of the molecule.

---

[8] A not directly observable quantity.

[9] The AOs may have *different* (as in H_2) or *identical* principal quantum numbers (as in LiH or Be$_2$, where 2s and 2p AOs are nearly degenerate). In H_2O, NH_3 and CH_4, 2s and 2p AOs are mixed under the C_{2v}, C_{3v} and T_d symmetry of the electric field provided by two, three, and four protons respectively.

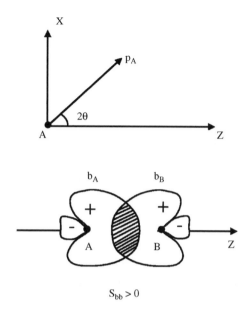

$$S_{bb} > 0$$

Figure 9.7 Overlap between two sp bond hybrids along the bond direction

- For the carbon atom, promotion of an electron from 2s to 2p *increases* its covalency from 2 to 4, giving the possibility of forming four equivalent sp^3 hybrids directed towards the vertices of a tetrahedron, making an interhybrid angle of 109.5°. The energy expended for promoting electrons is largely recovered by the formation of two additional strong single bonds.
- Hybridization increases overlap between the AOs forming the bond, yielding a stronger bond (more correctly, it increases the exchange-overlap component of the bond energy).

The interhybrid overlap between two sp hybrids b_A and b_B is shown in the bottom part of Figure 9.7. Assuming positive all elementary overlaps between s and pσ AOs, the overlap between two general sp hybrids making an angle 2θ with respect to the bond direction z is

$$S_{bb} = \langle b_A | b_B \rangle$$
$$= S_{ss} \cos^2\omega + (S_{\sigma\sigma} \cos^2 2\theta + S_{\pi\pi} \sin^2 2\theta)\sin^2\omega + S_{s\sigma} \cos 2\theta \sin 2\omega \tag{9.61}$$

Table 9.6 Overlap between sp bond hybrids along the bond direction

Hybrid	sp^n	$\cos \omega$	$\omega/°$	S_{bb}
Digonal	sp	$1/\sqrt{2}$	45	0.8017
Trigonal	sp^2	$1/\sqrt{3}$	54.7	0.7600
Tetrahedral	sp^3	$1/2$	60	0.7186

where ω is the hybridization parameter for *equivalent* hybrids; namely:

$$b_A = s_A\cos\omega + p_A\sin\omega, \quad b_B = s_B\cos\omega + p_B\sin\omega \tag{9.62}$$

with

$$p_A = \sigma_A\cos2\theta + \pi_A\sin2\theta \tag{9.63}$$

In (9.63), σ_A is a 2p AO directed along the bond and π_A is a 2p AO perpendicular to the bond direction.

For hybrids along the bond, $\theta = 0°$, so that (9.61) becomes

$$S_{bb}(\theta = 0°) = S_{ss}\cos^2\omega + S_{\sigma\sigma}\sin^2\omega + S_{s\sigma}\sin2\omega \tag{9.64}$$

For C=C in ethylene at $R = 2.55a_0$, the elementary overlaps between STOs with $c_s = c_p = 1.625$ are

$$S_{ss} = 0.4322, \quad S_{\sigma\sigma} = 0.3262, \quad S_{s\sigma} = 0.4225 \tag{9.65}$$

In Table 9.6 we give the overlap between equivalent hybrids for different values of the hybridization parameter ω. It is seen from the table that hybrid overlap is sensibly larger than each elementary overlap. Considering the largest overlap, $S_{ss} = 0.4322$, hybrid overlap is *increased* by over 85.5% for sp, 75.8% for sp^2 and 66.3% for sp^3. These numbers show the fantastic increase in overlap due to sp hybridization.

9.6 PAULING'S FORMULA FOR CONJUGATED AND AROMATIC HYDROCARBONS

To obtain energy values in VB theory it is necessary to evaluate matrix elements between structures, which may not be easy because the structures may not be orthogonal and the usual Slater rules for orthonormal determinants are not valid.

Pauling (1933) gave a simple pictorial formula for evaluating the matrix elements of the Hamiltonian for singlet covalent VB structures occurring in the theory of the π electrons of conjugated and aromatic hydrocarbons, which is of use in semiempirical calculations of the resonance energy much in the same way as is Hückel's theory we studied in Chapter 7. Based on the following assumptions:[10]

- orthonormality of the basic AOs
- consideration of singlet covalent structures only
- consideration of single interchanges between *adjacent* orbitals only

Pauling derived a simple formula for the matrix element of the Hamiltonian between the VB structures ψ_r and ψ_s as

$$H_{rs} - ES_{rs} = \frac{1}{2^{n-i}} \left(Q - E + \sum_{i,j} K_{ij} - \frac{1}{2} \sum_{i,j} K_{ij} \right) \qquad (9.66)$$

where n is the number of π bonds, i the number of islands in the superposition pattern, $Q \,(< 0)$ the Coulomb integral and $K_{ij} \,(< 0)$ the integral of single exchange between the pair i and j. The first summation in round brackets is over all *bonded* orbitals in the *same* island, the second over all *nonbonded* orbitals in *different* islands. The superposition patterns are closed polygons or *islands*, each formed by an even number of bonds, which are obtained by superposing the so-called Rumer *diagrams* describing single covalent bonds between singly occupied AOs, as depicted in Figure 9.8 for ethylene ($N = 2$), cyclobutadiene ($N = 4$), butadiene ($N = 4$), and the allyl radical ($N = 3$). The last molecule has an *odd* number of bonds, and is introduced just to show how Pauling's formula can be applied in this case.

It must be remarked that in the figures the σ skeleton is included just to recall the complete chemical structure of each molecule, but Rumer diagrams are concerned only with the π bonds (marked as double bonds in the figure). Above each superposition diagram is the corresponding coefficient obtained from Equation 9.66.[11]

---

[10] A detailed critique and partial justification of Pauling's hypotheses, which are not very different from those usually accepted for Hückel's theory, are fully discussed elsewhere (Magnasco, 2007).

[11] Because of molecular symmetry, all K_{ij} integrals are equal and denoted by K.

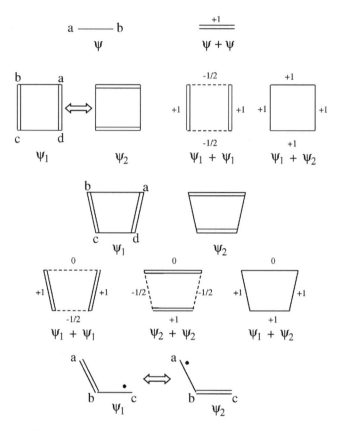

Figure 9.8 VB canonical structures and Pauling superposition patterns for ethylene, cyclobutadiene, butadiene and the allyl radical

As already done in Hückel's theory, in writing the VB secular equations it is convenient to put

$$\frac{Q-E}{K} = -x \Rightarrow E = Q + xK \qquad (9.67)$$

where x is the π bond energy in units of parameter K and $x > 0$ means *bonding*.

Paralleling what we have done for Hückel's theory in Chapter 7, we shall now apply Pauling's formula to ethylene, cyclobutadiene, butadiene, the allyl radical and benzene.

9.6.1 Ethylene (One π-Bond, $n = 1$)

We have $n = i = 1$, so that the (1×1) secular equation is

$$\begin{cases} H_{\psi\psi} - ES_{\psi\psi} = Q - E + K = 0 \\ |-x + 1| = 0 \Rightarrow x = 1 \Rightarrow E = Q + K \end{cases} \quad (9.68)$$

which is Pauling's energy of the *double* bond.

9.6.2 Cyclobutadiene ($n = 2$)

With reference to the second row of Figure 9.8, we can write for the two fully equivalent Kekulé structures[12] ψ_1 and ψ_2

$$\Psi = \psi_1 c_1 + \psi_2 c_2 \quad (9.69)$$

The distinct matrix elements are

$$\begin{cases} H_{11} - ES_{11} = H_{22} - ES_{22} = Q - E + 2K - K = Q - E + K \\ H_{12} - ES_{12} = H_{21} - ES_{21} = \frac{1}{2}(Q - E + 4K) = \frac{1}{2}(Q - E) + 2K \end{cases} \quad (9.70)$$

giving the (2×2) secular equation

$$\begin{vmatrix} -x + 1 & -\frac{x}{2} + 2 \\ -\frac{x}{2} + 2 & -x + 1 \end{vmatrix} = 0 \quad (9.71)$$

$$(-x + 1)^2 = \left(-\frac{x}{2} + 2\right)^2 \Rightarrow x^2 - 4 = 0 \Rightarrow x = \pm 2 \quad (9.72)$$

---

[12] Pauling VB theory distinguishes between Kekulé structures (all π bonds *equal*) and Dewar structures (containing one or more *long*, i.e. weaker, π bonds).

Taking the positive root, we have the following for the π energy of cyclobutadiene:

$$\begin{cases} 1 \text{ Kekulé}: & E = Q + K \\ 2 \text{ Kekulé}: & E = Q + 2K \end{cases} \qquad (9.73)$$

In Pauling's theory, the resonance energy is defined as the energy lowering resulting from the difference between the lowest root of the secular equation and the energy associated with the most stable single structure. In this case, denoting by K the Kekulé structure:

$$\Delta E = E(2K) - E(K) = K < 0 \qquad (9.74)$$

In this way, resonance between the two equivalent Kekulé structures *stabilizes* the system and the π energy decreases. This result is at variance from the corresponding Hückel result, where π delocalization energy was found to be zero for cyclobutadiene.

The calculation of the coefficients of the resonant VB functions in (9.69) proceeds as usual from the homogeneous system and the normalization condition:[13]

$$\begin{cases} (-x+1)c_1 + \left(-\dfrac{x}{2} + 2\right)c_2 + 0 \\ c_1^2 + c_2^2 = 1 \end{cases} \qquad (9.75)$$

$$c_2 = \frac{x-1}{2-\frac{x}{2}} \qquad c_1 = c_1 \quad \text{for } x = 2 \qquad (9.76)$$

so that we have

$$c_1 = c_2 = \frac{1}{\sqrt{2}} \qquad (9.77)$$

---

[13] For the sake of simplicity, unless differently stated, we assume orthogonality for the structures, not only for the AOs. This gives no problem at this elementary level.

giving

$$50\% \ \psi_1, \quad 50\% \ \psi_2 \tag{9.78}$$

as relative weights of the two Kekulé structures in square cyclobutadiene, as it must be for completely resonant structures.

9.6.3   Butadiene (Open Chain, n = 2)

Butadiene, the *open* chain with $2n = 4$ (two π bonds), has the canonical structures and the superposition patterns depicted in the third and fourth rows of Figure 9.8. ψ_1 is the Kekulé structure (two equivalent *short* π bonds) and ψ_2 is the Dewar structure (one *long* π bond). The complete VB wavefunction will be

$$\Psi = \psi_1 \, c_1 + \psi_2 \, c_2 \tag{9.79}$$

where the coefficients must be determined by the Ritz method.

The matrix elements are

$$\begin{cases} H_{11} - ES_{11} = Q - E + \dfrac{3}{2} K \\[2mm] H_{22} - ES_{22} = Q - E \\[2mm] H_{12} - ES_{12} = H_{21} - ES_{21} = \dfrac{1}{2}(Q - E + 3K) = \dfrac{1}{2}(Q - E) + \dfrac{3}{2} K \end{cases} \tag{9.80}$$

The (2×2) secular equation is

$$\begin{vmatrix} -x + \dfrac{3}{2} & -\dfrac{x}{2} + \dfrac{3}{2} \\[3mm] -\dfrac{x}{2} + \dfrac{3}{2} & -x \end{vmatrix} = 0 \tag{9.81}$$

$$-x\left(-x + \dfrac{3}{2}\right) = \left(-\dfrac{x}{2} + \dfrac{3}{2}\right)^2 \Rightarrow x^2 - 3 = 0 \Rightarrow x = \pm\sqrt{3} \tag{9.82}$$

Taking the positive root, we have the following for the π energy of butadiene:

$$\begin{cases} 1 \text{ Kekulé}: \qquad\qquad E = Q + \dfrac{3}{2}K \\[2mm] 1 \text{ Kekulé} + 1 \text{ Dewar}: \quad E = Q + \sqrt{3}K \end{cases} \qquad (9.83)$$

so that we obtain

$$\Delta E = E(K+D) - E(K) = (1.73 - 1.5)K = 0.23K < 0 \qquad (9.84)$$

for the *conjugation energy*.[14]

Calculation of the coefficients in butadiene proceeds through solution of the system:

$$\begin{cases} \left(-x + \dfrac{3}{2}\right)c_1 + \left(-\dfrac{x}{2} + \dfrac{3}{2}\right)c_2 + 0 \\[2mm] c_1^2 + c_2^2 = 1 \end{cases} \qquad (9.85)$$

$$c_2 = \frac{x - \frac{3}{2}}{\frac{3}{2} - \frac{x}{2}}c_1 = 0.362c_1 \quad \text{for } x = \sqrt{3} \approx 1.73 \qquad (9.86)$$

Therefore, we obtain the following for the 'resonance' between Kekulé and Dewar structures in butadiene:

$$\Psi = c_1(\psi_1 + 0.362\,\psi_2) = 0.94\,\psi_1 + 0.34\,\psi_2 \qquad (9.87)$$

giving

$$88\,\%\,\psi_1, \quad 12\,\%\,\psi_2 \qquad (9.88)$$

which shows the greater importance of the Kekulé structure versus the Dewar structure.[15]

---

[14] Since the structures are *different*, it is appropriate to speak of conjugation and not of resonance energy.

[15] Always, Dewar structures, having *long* bonds, are sensibly less important than Kekulé structures, but their importance may increase with their number (e.g. naphthalene and anthracene).

9.6.4 The Allyl Radical (N = 3)

This case (last row of Figure 9.8) is an interesting example of how to apply Pauling's rules to an *odd-electron* system ($N = 3$, $S = \frac{1}{2}$). We add a *phantom* atom (say d), treat the system as a four-atom system and, at the end, *remove* the contribution of the phantom atom from the calculation. The covalent VB structures and the corresponding superposition patterns are then the same as those of butadiene. Hence, with reference to the previous calculation, we can write

$$
\begin{cases}
H_{11} - E S_{11} = Q - E + K - \dfrac{1}{2}K + \mathbf{K} = Q - E + \dfrac{1}{2}K \\[2mm]
H_{22} - E S_{22} = Q - E + K - \dfrac{1}{2}K - \dfrac{1}{2}\mathbf{K} = Q - E + \dfrac{1}{2}K \\[2mm]
H_{12} - E S_{12} = H_{21} - E S_{21} = \dfrac{1}{2}(Q - E + K + K + \mathbf{K}) = \dfrac{1}{2}(Q - E) + K
\end{cases}
$$

$$(9.89)$$

where we have bolded the contributions which must be removed. We then obtain the secular equation for the allyl radical:

$$
\begin{vmatrix}
-x + \dfrac{1}{2} & -\dfrac{x}{2} + 1 \\[3mm]
-\dfrac{x}{2} + 1 & -x + \dfrac{1}{2}
\end{vmatrix} = 0
\qquad (9.90)
$$

$$
\left(-x + \frac{1}{2}\right)^2 = \left(-\frac{x}{2} + 1\right)^2 \Rightarrow \frac{3}{4}x^2 - \frac{3}{4} = 0 \Rightarrow x = \pm 1
\qquad (9.91)
$$

Taking the positive root, we have the following for the π energy of the radical:

$$
\begin{cases}
1\ \text{Kekulé}: & E = Q + \dfrac{1}{2}K \\[2mm]
1\ \text{Kekulé} + 1\ \text{Dewar}: & E = Q + K
\end{cases}
\qquad (9.92)
$$

so that we obtain

$$\Delta E = E(K+D) - E(K) = \frac{1}{2}K < 0 \qquad (9.93)$$

for the *conjugation energy*. The conjugation energy in the allyl radical is hence *larger* than that of butadiene (nearly twice).

Calculation of the coefficients in the allyl radical proceeds through solution of the system

$$\begin{cases} \left(-x+\frac{1}{2}\right)c_1 + \left(-\frac{x}{2}+1\right)c_2 + 0 \\ c_1^2 + c_2^2 = 1 \end{cases} \qquad (9.94)$$

$$c_2 = \frac{x-\frac{1}{2}}{1-\frac{x}{2}}c_1 = c_1 \quad \text{for } x = 1 \qquad (9.95)$$

giving

$$c_1 = c_2 = \frac{1}{\sqrt{2}} \qquad (9.96)$$

if we neglect nonorthogonality between the structures.

Therefore, the relative weights of the two structures ψ_1 and ψ_2 are

$$50\,\%\ \psi_1, \quad 50\,\%\ \psi_2 \qquad (9.97)$$

as it must be for two truly *resonant* structures (ψ_1 and ψ_2 are fully equivalent).

By writing the complete VB wavefunctions for the two resonant structures, we can immediately obtain electron and spin density distributions in the allyl radical, even without doing any effective calculation of the energy. Assuming orthonormal spin-orbitals, with reference to Figure 9.8, the complete form of the two structures in terms of their parents is

$$\psi_1 = \frac{1}{\sqrt{2}}\left[(a\bar{b}c) - (\bar{a}bc)\right] \qquad (9.98)$$

$$\psi_2 = \frac{1}{\sqrt{2}}\left[(a\bar{b}c)-(ab\bar{c})\right] \tag{9.99}$$

The structures have the *same* parent but *different* wavefunctions. Even if the spin-orbitals are assumed orthonormal, the structures are non-orthogonal (they have equal the first determinant):

$$S_{12} = \langle\psi_1|\psi_2\rangle = \frac{1}{2} \tag{9.100}$$

Then, from

$$\Psi = \psi_1 c_1 + \psi_2 c_2 \tag{9.101}$$

it is immediately obtained that

$$c_1 = c_2 = \frac{1}{\sqrt{3}} \tag{9.102}$$

The apparently strange normalization of the wavefunction (9.101) is due to the nonorthogonality of ψ_1 and ψ_2, as can be easily checked:

$$\langle\Psi|\Psi\rangle = c_1^2 + c_2^2 + 2c_1 c_2 S_{12} = \frac{1}{3} + \frac{1}{3} + 2 \times \frac{1}{3} \times \frac{1}{2} = 1 \tag{9.103}$$

Direct calculation of the α- and β-components of the one-electron density gives

$\rho_1^\alpha = $ coefficient of $\alpha\alpha^*$ in ρ_1

$$= a^2\left(\frac{1}{2}c_1^2 + c_2^2 + c_1 c_2\right) + b^2\left(\frac{1}{2}c_1^2 + \frac{1}{2}c_2^2\right) + c^2\left(c_1^2 + \frac{1}{2}c_2^2 + c_1 c_2\right) \tag{9.104}$$

$\rho_1^\beta = $ coefficient of $\beta\beta^*$ in ρ_1

$$= a^2\left(\frac{1}{2}c_1^2\right) + b^2\left(\frac{1}{2}c_1^2 + \frac{1}{2}c_2^2 + c_1 c_2\right) + c^2\left(\frac{1}{2}c_2^2\right) \tag{9.105}$$

Therefore, we have

$$P(\mathbf{r}) = \rho_1^\alpha(\mathbf{r}) + \rho_1^\beta(\mathbf{r})$$
$$= (c_1^2 + c_2^2 + c_1 c_2)(a^2 + b^2 + c^2) = a^2 + b^2 + c^2 \tag{9.106}$$

$$Q(\mathbf{r}) = \rho_1^\alpha(\mathbf{r}) - \rho_1^\beta(\mathbf{r})$$

$$= a^2(c_1^2 + c_1 c_2) - b^2(c_1 c_2) + c^2(c_1^2 + c_1 c_2) = \frac{2}{3}a^2 - \frac{1}{3}b^2 + \frac{2}{3}c^2$$

$$\tag{9.107}$$

So, while the π electron density (9.106) is uniform (one electron onto each carbon atom, as it must be for an alternant hydrocarbon), VB theory predicts that an excess of α spin at the terminal atoms induces some β spin at the central atom, in accord with experimental ESR results. This is an interesting example of electron and spin population analysis in *multideterminant* wavefunctions.

9.6.5   Benzene (n = 3)

The five canonical structures (2 Kekulé + 3 Dewar) and their distinct superposition patterns for the six π electrons of benzene are given in Figure 9.9. The matrix elements are

$$\begin{cases} H_{11} - ES_{11} = H_{22} - ES_{22} = Q - E + \dfrac{3}{2}K \\[2mm] H_{12} - ES_{12} = \dfrac{1}{4}(Q - E) + \dfrac{3}{2}K \end{cases} \tag{9.108}$$

$$\begin{cases} H_{33} - ES_{33} = H_{44} - ES_{44} = H_{55} - ES_{55} = Q - E \\[2mm] H_{34} - ES_{34} = H_{35} - ES_{35} = H_{45} - ES_{45} = \dfrac{1}{4}(Q - E) + \dfrac{3}{2}K \end{cases} \tag{9.109}$$

$$\begin{aligned} H_{13} - ES_{13} = H_{14} - ES_{14} &= H_{15} - ES_{15} \\ &= H_{23} - ES_{23} = H_{24} - ES_{24} = H_{25} - ES_{25} \\ &= \frac{1}{2}(Q - E) + \frac{3}{2}K \end{aligned} \tag{9.110}$$

We now examine the solution of the secular equations in three different cases.

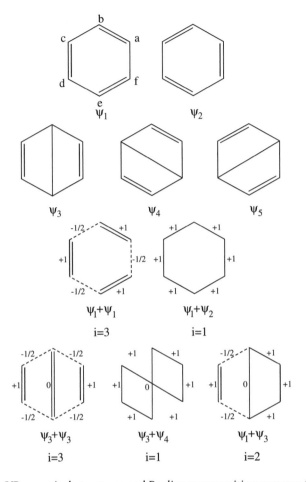

Figure 9.9 VB canonical structures and Pauling superposition patterns for benzene

- *Resonance between the two Kekulé structures*

$$\Psi = \psi_1 c_1 + \psi_2 c_2 \tag{9.111}$$

$$\begin{vmatrix} -x + \dfrac{3}{2} & -\dfrac{x}{4} + \dfrac{3}{2} \\[2mm] -\dfrac{x}{4} + \dfrac{3}{2} & -x + \dfrac{3}{2} \end{vmatrix} = 0 \tag{9.112}$$

$$5x^2 - 12x = x(5x - 12) = 0 \Rightarrow x = 0, \ x = \frac{12}{5} \tag{9.113}$$

Taking the lowest root, we obtain for the ground state

$$E = Q + \frac{12}{5}K = Q + 2.4K \qquad (9.114)$$

giving as *resonance energy* for the two Kekulé structures

$$\Delta E = E(2K) - E(K) = \left(\frac{12}{5} - \frac{3}{2}\right)K = \frac{9}{10}K < 0 \qquad (9.115)$$

- **Resonance between the three Dewar structures**

$$\Psi = \psi_3 c_3 + \psi_4 c_4 + \psi_5 c_5 \qquad (9.116)$$

$$\begin{vmatrix} -x & -\dfrac{x}{4}+\dfrac{3}{2} & -\dfrac{x}{4}+\dfrac{3}{2} \\[2ex] -\dfrac{x}{4}+\dfrac{3}{2} & -x & -\dfrac{x}{4}+\dfrac{3}{2} \\[2ex] -\dfrac{x}{4}+\dfrac{3}{2} & -\dfrac{x}{4}+\dfrac{3}{2} & -x \end{vmatrix} = 0 \qquad (9.117)$$

Expanding the determinant gives the cubic equation in x:

$$x^3 + 2x^2 - 4x - 8 = (x+2)^2(x-2) = 0 \Rightarrow x = -2 \text{ (twice)}, \ x = 2 \qquad (9.118)$$

The lowest root is

$$E = Q + 2K \qquad (9.119)$$

giving as *resonance energy* between the three Dewar structures

$$\Delta E = E(3D) - E(K) = \left(2 - \frac{3}{2}\right)K = \frac{1}{2}K < 0 \qquad (9.120)$$

The resonance between the three Dewar structures, each having a *long* bond, is therefore sensibly less than the resonance between the two Kekulé structures (0.5 instead of 0.9).

• *Resonance between all VB structures*

The complete VB problem, arising from the mixing of all five VB structures, is rather tedious since it would involve solution of a fifth-order determinantal equation. We can, however, simplify the problem using symmetry arguments, if we are only interested in the ground-state energy of the system, as we are. In fact, symmetry suggests that

$$\Psi = (\psi_1 + \psi_2)c_K + (\psi_3 + \psi_4 + \psi_5)c_D = \Psi_K c_K + \Psi_D c_D \qquad (9.121)$$

where Ψ_K and Ψ_D are the un-normalized combinations of *equivalent* Kekulé and Dewar structures respectively. In this way, we reduce the full-VB problem to the solution of a simple quadratic secular equation. We have

$$\begin{vmatrix} H_{KK} - ES_{KK} & H_{KD} - ES_{KD} \\ H_{KD} - ES_{KD} & H_{DD} - ES_{DD} \end{vmatrix} = 0 \qquad (9.122)$$

where, from (9.108)–(9.110):

$$\begin{cases} H_{KK} - ES_{KK} = \dfrac{5}{2}(Q - E) + 6K, \quad H_{DD} - ES_{DD} = \dfrac{9}{2}(Q - E) + 9K \\ H_{KD} - ES_{KD} = 3(Q - E) + 9K \end{cases}$$
$$(9.123)$$

Therefore, the (2×2) secular equation is

$$\begin{vmatrix} -\dfrac{5}{2}x + 6 & -3x + 9 \\ -3x + 9 & -\dfrac{9}{2}x + 9 \end{vmatrix} = 0 \qquad (9.124)$$

giving upon expansion

$$x^2 + 2x - 12 = 0 \Rightarrow x = -1 + \sqrt{13}, \ x = -1 - \sqrt{13} \qquad (9.125)$$

Taking the lowest root, we obtain the strongly bonding ground state

$$E = Q + 2.6055K \qquad (9.126)$$

with the resonance energy

$$\Delta E = E(2K + 3D) - E(K) = (2.6055 - 1.5)K = 1.1K \qquad (9.127)$$

which is the largest seen so far for benzene.

For the coefficients of the resonant structures, we have from the homogeneous system

$$c_D = \frac{\frac{5}{2}x - 6}{-3x + 9} c_K = 0.4321 c_K \quad \text{for } x = 2.6055 \qquad (9.128)$$

Assuming orthogonality between the structures, we obtain the normalization factor for the Ψ in (9.121):

$$\langle \Psi | \Psi \rangle = 2c_K^2 + 3c_D^2 = 1 \Rightarrow N = (2c_K^2 + 3c_D^2)^{-1/2} \qquad (9.129)$$

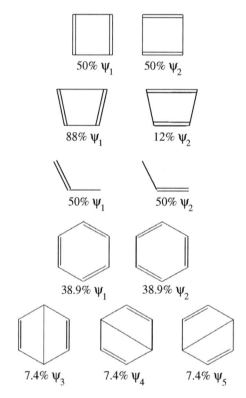

Figure 9.10   Percentage relative weights of VB canonical structures in conjugated and aromatic hydrocarbons

$$c_K = 0.6243, \quad c_D = 0.2710 \tag{9.130}$$

finally giving the following as relative weights of the covalent singlet structures in benzene:

$$38.9\,\%\,\psi_1, \quad 38.9\,\%\,\psi_2, \quad 7.4\,\%\,\psi_3, \quad 7.4\,\%\,\psi_4, \quad 7.4\,\%\,\psi_5$$
$$\tag{9.131}$$

So, the contribution of the Dewar structures, even if individually small, is quite important on the whole. The results (9.131) are in surprising agreement with recent ab initio VB calculations by Cooper *et al.* (1986), which give for these weights 40.3% for $\psi_{1,2}$ and 6.5% for ψ_{3-5}.

For the sake of comparison, the relative weights of the structures resulting from all preceding VB calculations are collected in Figure 9.10.

10

Elements of Rayleigh–Schroedinger Perturbation Theory

In this chapter we present a few elements of RS *stationary* perturbation theory, where a stationary system acted upon by an external perturbation will change by a small amount its energy levels and eigenstates. It consists essentially in relating the actual eigenvalue problem to be solved to one for which a complete solution is exactly known, and in treating the difference between the two Hamiltonian operators as a small perturbation.

10.1 RAYLEIGH–SCHROEDINGER PERTURBATION EQUATIONS UP TO THIRD ORDER

We want to solve the Schroedinger eigenvalue equation

$$(\hat{H}-E)\psi = 0 \qquad (10.1)$$

for a *Hermitian* decomposition of the Hamiltonian $\hat{H}$ into:

$$\hat{H} = \hat{H}_0 + \lambda \hat{H}_1 \qquad (10.2)$$

where (i) λ is a parameter giving the *orders* in perturbation theory,[1] (ii) $\hat{H}_0$ is the *unperturbed* Hamiltonian, namely the Hamiltonian of an already

---

[1] Orders are here carried by the suffixes or their sum.

Methods of Molecular Quantum Mechanics: An Introduction to Electronic Molecular Structure
Valerio Magnasco
© 2009 John Wiley & Sons, Ltd

solved problem (either physical or model) and (iii) $\hat{H}_1$ is the *small* first-order difference between $\hat{H}$ and $\hat{H}_0$, called the *perturbation*.

We now expand both eigenvalue E and eigenfunction ψ into powers of λ:

$$E = E_0 + \lambda E_1 + \lambda^2 E_2 + \lambda^3 E_3 + \cdots \qquad (10.3)$$

$$\psi = \psi_0 + \lambda \psi_1 + \cdots \qquad (10.4)$$

where the coefficients of the different powers of λ are, respectively, the *corrections* of the various orders to energy and wavefunction (e.g. E_2 is the second-order energy correction, ψ_1 the first-order correction to the wavefunction, and so on). It is often useful to define corrections up to a given order, which we write, for example, as

$$E^{(3)} = E_0 + E_1 + E_2 + E_3 \qquad (10.5)$$

meaning that we add corrections *up to* the third order.

By substituting the expansions into the Schroedinger Equation 10.1:

$$[(\hat{H}_0 - E_0) + \lambda(\hat{H}_1 - E_1) - \lambda^2 E_2 - \lambda^3 E_3 - \cdots](\psi_0 + \lambda\psi_1 + \lambda^2\psi_2 + \cdots) = 0 \qquad (10.6)$$

and separating orders, we obtain

$$\begin{cases} \lambda^0 & (\hat{H}_0 - E_0)\psi_0 = 0 \\ \lambda & (\hat{H}_0 - E_0)\psi_1 + (\hat{H}_1 - E_1)\psi_0 = 0 \\ \lambda^2 & (\hat{H}_0 - E_0)\psi_2 + (\hat{H}_1 - E_1)\psi_1 - E_2\psi_0 = 0 \\ \cdots \end{cases} \qquad (10.7)$$

which are known as RS perturbation equations of the various orders specified by the power of λ.

Because of the Hermitian property of $\hat{H}_0$, bracketing Equations 10.7 on the left by $\langle\psi_0|$, all the first terms in the RS equations are zero, and we are left with

$$\begin{cases} \lambda^0 & \langle\psi_0|\hat{H}_0 - E_0|\psi_0\rangle = 0 \\ \lambda & \langle\psi_0|\hat{H}_1 - E_1|\psi_0\rangle = 0 \\ \lambda^2 & \langle\psi_0|\hat{H}_1 - E_1|\psi_1\rangle - E_2\langle\psi_0|\psi_0\rangle = 0 \\ \cdots \end{cases} \qquad (10.8)$$

Taking ψ_0 normalized to 1, we obtain the RS energy corrections of the various orders as

$$
\begin{cases}
\lambda^0 & E_0 = \langle \psi_0 | \hat{H}_0 | \psi_0 \rangle \\
\lambda & E_1 = \langle \psi_0 | \hat{H}_1 | \psi_0 \rangle \\
\lambda^2 & E_2 = \langle \psi_0 | \hat{H}_1 - E_1 | \psi_1 \rangle = -\langle \psi_1 | \hat{H}_0 - E_0 | \psi_1 \rangle \\
\lambda^3 & E_3 = \langle \psi_0 | \hat{H}_1 - E_1 | \psi_2 \rangle - E_2 \langle \psi_0 | \psi_1 \rangle = \langle \psi_1 | \hat{H}_1 - E_1 | \psi_1 \rangle \\
& \cdots
\end{cases}
\tag{10.9}
$$

E_0 and E_1 are nothing but the average values of $\hat{H}_0$ and $\hat{H}_1$ respectively over the unperturbed function ψ_0, while E_2 is given as a *nondiagonal* term, often referred to as *transition integral*, connecting ψ_0 to ψ_1 through the operator $\hat{H}_1$, the last expression of E_2 in (10.9) showing that always $E_2 < 0$. The equations above show that knowledge of ψ_1 (the solution of the first-order RS differential equation) determines the energy corrections up to third order.[2] In solving the first-order RS differential equation, we impose on ψ_1 the *orthogonality* condition

$$
\langle \psi_0 | \psi_1 \rangle = 0
\tag{10.10}
$$

which follows in first order from the normalization condition on the total wavefunction:

$$
\langle \psi | \psi \rangle = 1
\tag{10.11}
$$

and the fact that we assume a normalized ψ_0:

$$
\langle \psi_0 | \psi_0 \rangle = 1
\tag{10.12}
$$

We recall that all ψ_n $(n \neq 0)$ corrections are *not* normalized (even if they are normalizable).

Before proceeding any further, an explanation is due to the reader of the symmetric form resulting for E_3 in (10.9). In fact, the RS third-order differential equation determining ψ_3:

$$
\lambda^3 (\hat{H}_0 - E_0)\psi_3 + (\hat{H}_1 - E_1)\psi_2 - E_2\psi_1 - E_3\psi_0 = 0
\tag{10.13}
$$

---

[2] In general, ψ_n determines E up to E_{2n+1}.

gives the third-order energy correction in the form

$$\lambda^3 E_3 = \langle \psi_0 | \hat{H}_1 - E_1 | \psi_2 \rangle - E_2 \langle \psi_0 | \psi_1 \rangle \qquad (10.14)$$

It will now be shown that it is possible to shift the order from the operator to the wavefunction, and vice versa,[3] if we make repeated use of the RS perturbation equations given in (10.7) and take into account the fact that the operators are Hermitian. In fact, we can write

$$E_3 = \langle (\hat{H}_1 - E_1) \psi_0 | \psi_2 \rangle - E_2 \langle \psi_0 | \psi_1 \rangle \qquad (10.15)$$

and, using the complex conjugate of the first-order equation (the bra):

$$\begin{aligned} E_3 &= -\langle (\hat{H}_0 - E_0) \psi_1 | \psi_2 \rangle - E_2 \langle \psi_0 | \psi_1 \rangle \\ &= -\langle \psi_1 | \hat{H}_0 - E_0 | \psi_2 \rangle - E_2 \langle \psi_0 | \psi_1 \rangle \end{aligned} \qquad (10.16)$$

In this equation, the order has been shifted from the operator to the wavefunction. If we now make use of the second-order RS differential equation, the last term above can be written

$$E_3 = \langle \psi_1 | \hat{H}_1 - E_1 | \psi_1 \rangle - E_2 [\langle \psi_0 | \psi_1 \rangle + \langle \psi_1 | \psi_0 \rangle] \qquad (10.17)$$

Since the term in square brackets is identically zero, the last of Equations 10.9 is recovered. The same can be done for E_2.

Finally, it must be emphasized that the leading term of the RS perturbation equations (10.7), the zeroth-order equation $(\hat{H}_0 - E_0)\psi_0 = 0$, must be satisfied *exactly*, otherwise uncontrollable errors will affect the whole chain of equations. Furthermore, it must be observed that only energy in first order gives an upper bound to the true energy of the ground state, so that the energy in second order, $E^{(2)}$, may be *below* the true value.[4]

10.2   FIRST-ORDER THEORY

First-order RS theory is useful, for instance, in explaining the Zeeman effect and the splitting of the multiplet structure in atoms, or in giving the

---

[3] This is known as the Dalgarno interchange theorem.
[4] This is particularly true when the correct value of E_2 is determined.

Coulombic component of the interaction energy between atoms and molecules.

An interesting case arises when the zeroth-order energy level E_0 is N-fold degenerate. In this case, it can be shown that the first-order energy corrections are given by the solutions of the first-order determinantal equation of degree N:

$$|\mathbf{H}_1 - E_1 \mathbf{1}| = 0 \tag{10.18}$$

where $\mathbf{H}_1$ is the representative of the perturbation $\hat{H}_1$ over the orthonormal set $\{\psi_1^0, \psi_2^0, \ldots, \psi_N^0\}$ belonging to the degenerate eigenvalue E_0. A striking example of this first-order degenerate perturbation theory is offered by the Hückel theory of chain hydrocarbons. In this case, the determinantal equation (10.18) is nothing but the Hückel determinant D_N:

$$D_N = \begin{vmatrix} -x & 1 & 0 & \cdots \\ 1 & -x & 1 & \cdots \\ \cdots & \cdots & \cdots & \cdots \\ \cdots & 0 & 1 & -x \end{vmatrix} = 0 \tag{10.19}$$

whose solutions were studied in Chapter 7.

10.3   SECOND-ORDER THEORY

Apart from first-order theory, essential to any practical use of perturbation theory is the possibility of solving, at least, the first-order RS differential equation. This can be done exactly in some cases, as for the H atom in a uniform electric field.

The problem is best understood in its physical terms if we use the simple model described in detail elsewhere (Magnasco, 2004b). The H atom, spherical in its ground state, *distorts* under the action of a uniform external field F ($F_x = F_y = 0$, $F_z = F$) acquiring an *induced* dipole moment proportional to the field F:

$$\mu_i = \alpha F \tag{10.20}$$

where α is a second-order electric property of the atom, called the (dipole) *polarizability*.[5] We define α (Figure 10.1) in terms of the (dipole)

---

[5] The polarizability is the physical quantity measuring the distortion of the atom (or molecule) under the action of the field; it has the dimensions of a volume, its atomic units being a_0^3.

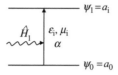

Figure 10.1   Dipole transition between ψ_0 and ψ_1 for ground-state H

transition moment μ_i from state ψ_0 to ψ_1 and its related excitation energy $\varepsilon_i \, (> 0)$ as

$$\alpha = \frac{2\mu_i^2}{\varepsilon_i} \tag{10.21}$$

In the uniform external field F, the atom acquires the potential energy

$$\hat{H}_1 = -Fz \tag{10.22}$$

which will be taken as the first-order perturbation. $\lambda = F$ is now a *physical* perturbation parameter which can be acted upon from the laboratory (e.g. by swithching on/off a plane condenser).

A simple Ritz variational calculation involving ψ_0 and ψ_1, both states being assumed normalized and orthogonal to each other, shows that the energy of the atom in the field is lowered by the amount

$$
\begin{aligned}
\Delta E &= -\frac{|\langle \psi_0 | \hat{H}_1 | \psi_1 \rangle|^2}{\varepsilon_i} = -F^2 \frac{(\psi_0 \psi_1 | z)^2}{\varepsilon_i} \\
&= -F^2 \frac{\mu_i^2}{\varepsilon_i} = -\frac{1}{2}\left(\frac{2\mu_i^2}{\varepsilon_i}\right) F^2 = -\frac{1}{2}\alpha F^2
\end{aligned}
\tag{10.23}
$$

From this relation follows that α can be obtained as the negative of the second derivative of the energy with respect to the field F evaluated at $F = 0$:

$$\alpha = -\left(\frac{\mathrm{d}^2 \Delta E}{\mathrm{d}F^2}\right)_{F=0} \tag{10.24}$$

Turning to our perturbation theory, we take

$$\hat{H} = \hat{H}_0 - Fz, \quad \hat{H}_0 = -\frac{1}{2}\nabla^2 - \frac{1}{r} \tag{10.25}$$

where $z = r\cos\theta$. Then, the first-order energy correction is zero by symmetry:

$$E_1 = \langle\psi_0|-Fz|\psi_0\rangle = -F\langle\psi_0|z|\psi_0\rangle = 0 \qquad (10.26)$$

since z is *odd* and the spherical density $\{\psi_0\}^2$ *even*. In this case, the first physically relevant correction to the energy occurs in second order of perturbation theory. It can be shown (see Magnasco (2007)) that the *exact* first-order dipole correction to the unperturbed wavefunction ψ_0 is

$$\psi_1 = F\left(r + \frac{r^2}{2}\right)\psi_0\cos\theta \qquad (10.27)$$

where

$$\psi_0 = \frac{1}{\sqrt{\pi}}\exp(-r) \qquad (10.28)$$

is the unperturbed wavefunction for the ground state of the H atom. The student can verify this solution taking into account that

$$\nabla^2 = \nabla_r^2 - \frac{\hat{L}^2}{r^2}, \quad \nabla_r^2 = \frac{d^2}{dr^2} + \frac{2}{r}\frac{d}{dr}, \quad \hat{L}^2\cos\theta = 2\cos\theta, \quad E_0 = -\frac{1}{2}$$

$$(10.29)$$

The *exact* second-order correction to the energy is then given by

$$E_2 = \langle\psi_0|\hat{H}_1|\psi_1\rangle = \langle\psi_0|-Fz|\psi_1\rangle$$

$$= -F^2\left\langle\psi_0\left|z\left(r + \frac{r^2}{2}\right)\cos\theta\right|\psi_0\right\rangle = -F^2\frac{9}{4} \qquad (10.30)$$

where the integral is easily evaluated in spherical coordinates ($x = \cos\theta$):

$$\left\langle\psi_0\left|\left(r^2 + \frac{r^3}{2}\right)\cos^2\theta\right|\psi_0\right\rangle = 2\pi\frac{1}{\pi}\int_{-1}^{1}dx\,x^2\int_0^\infty dr\,r^2\left(r^2 + \frac{r^3}{2}\right)\exp(-2r)$$

$$= \frac{4}{3}\int_0^\infty dr\left(r^4 + \frac{r^5}{2}\right)\exp(-2r) = \frac{4}{3}\left(\frac{4\times3\times2}{2^5} + \frac{15\times4\times3\times2}{2^6}\right) = \frac{9}{4}$$

$$(10.31)$$

The *exact* dipole polarizability (atomic units) of ground-state H is then obtained as

$$\alpha = -\left(\frac{d^2 E_2}{dF^2}\right)_0 = \frac{9}{2} = 4.5 a_0^3 \tag{10.32}$$

10.4 APPROXIMATE E_2 CALCULATIONS: THE HYLLERAAS FUNCTIONAL

When the first-order RS differential equation cannot be solved exactly, use must be made of the second-order Hylleraas functional:

$$\tilde{E}_2[\tilde{\psi}_1] = \langle \tilde{\psi}_1 | \hat{H}_0 - E_0 | \tilde{\psi}_1 \rangle + \langle \tilde{\psi}_1 | \hat{H}_1 - E_1 | \psi_0 \rangle + \langle \psi_0 | \hat{H}_1 - E_1 | \tilde{\psi}_1 \rangle \tag{10.33}$$

This gives the *exact* E_2 for the *exact* ψ_1, as can be easily verified by direct substitution in (10.33). In fact:

$$\begin{aligned} \tilde{E}_2[\psi_1] &= \langle \psi_1 | (\hat{H}_0 - E_0)\psi_1 + (\hat{H}_1 - E_1)\psi_0 \rangle + \langle \psi_0 | \hat{H}_1 - E_1 | \psi_1 \rangle \\ &= \langle \psi_0 | \hat{H}_1 - E_1 | \psi_1 \rangle = E_2 \end{aligned} \tag{10.34}$$

since the first term on the right-hand side vanishes, because the ket is identically zero when ψ_1 is the exact solution of the first-order equation. For *approximate* ψ_1, say $\tilde{\psi}_1$, it can be shown that we obtain *upper bounds* to the true E_2:

$$\tilde{E}_2[\tilde{\psi}_1] = \tilde{E}_2 \geq E_2 \tag{10.35}$$

and that the variational $\tilde{E}_2$ is affected by a *second-order* error.[6] In this way, we can easily construct *variational approximations* to second-order energies and first-order wavefunctions. Much as we did for the total energy in Chapter 4, we introduce variational (linear or nonlinear) parameters into $\tilde{\psi}_1$ and *minimize* the Hylleraas functional $\tilde{E}_2$ with respect to them, obtaining in this way the *best* variational approximation to $\tilde{E}_2$ and $\tilde{\psi}_1$.

---

[6] The wavefunction is instead affected by a *first-order* (i.e. larger) error. As in all variational approximations, energy is determined better than the wavefunction.

10.5 LINEAR PSEUDOSTATES AND MOLECULAR PROPERTIES

A convenient way to proceed is to apply the Ritz method to $\tilde{E}_2$. We start from a convenient set of basis functions χ written as the $(1 \times N)$ row vector:

$$\chi = (\chi_1 \chi_2 \cdots \chi_N) \tag{10.36}$$

possibly orthonormal in themselves but necessarily *orthogonal* to ψ_0. We shall assume that

$$\chi^\dagger \chi = 1, \quad \chi^\dagger \psi_0 = 0 \tag{10.37}$$

If the χs are not orthogonal then they must be preliminarily orthogonalized by the Schmidt method. Then, we construct the matrices

$$\mathbf{M} = \chi^\dagger (\hat{H}_0 - E_0) \chi \tag{10.38}$$

the $(N \times N)$ Hermitian matrix of the *excitation energies*, and

$$\mu = \chi^\dagger (\hat{H}_1 \psi_0) \tag{10.39}$$

the $(N \times 1)$ column vector of the transition moments.

By expanding ψ_1 in the *finite* set of the χs, we can write

$$\psi_1 = \chi \mathbf{C} = \sum_{\kappa=1}^{N} \chi_\kappa C_\kappa \tag{10.40}$$

$$\tilde{E}_2 = \mathbf{C}^\dagger \mathbf{M} \mathbf{C} + \mathbf{C}^\dagger \mu + \mu^\dagger \mathbf{C} \tag{10.41}$$

which is minimum for

$$\frac{\delta \tilde{E}_2}{\delta \mathbf{C}^\dagger} = \mathbf{M} \mathbf{C} + \mu = 0 \Rightarrow \mathbf{C} \,(\text{best}) = -\mathbf{M}^{-1} \mu \tag{10.42}$$

giving as *best* variational approximation to $\tilde{E}_2$

$$\tilde{E}_2 \,(\text{best}) = -\mu^\dagger \mathbf{M}^{-1} \mu \tag{10.43}$$

The Hermitian matrix $\mathbf{M}$ can be reduced to *diagonal* form by a unitary transformation $\mathbf{U}$ among its basis functions $\boldsymbol{\chi}$:

$$\psi = \boldsymbol{\chi}\mathbf{U}, \quad \mathbf{U}^\dagger\mathbf{M}\mathbf{U} = \boldsymbol{\varepsilon}, \quad \mathbf{U}^\dagger\boldsymbol{\mu} = \boldsymbol{\mu}_\psi, \qquad (10.44)$$

where $\boldsymbol{\varepsilon}$ is here the $(N \times N)$ diagonal matrix of the (positive) *excitation energies*:

$$\boldsymbol{\varepsilon} = \begin{pmatrix} \varepsilon_1 & 0 & \cdots & 0 \\ 0 & \varepsilon_2 & \cdots & 0 \\ \cdots & \cdots & \cdots & \cdots \\ 0 & 0 & \cdots & \varepsilon_N \end{pmatrix} \qquad (10.45)$$

The ψs are called *pseudostates* and give $\tilde{E}_2$ in the form

$$\tilde{E}_2 \,(\text{best}) = -\boldsymbol{\mu}_\psi^\dagger \boldsymbol{\varepsilon}^{-1} \boldsymbol{\mu}_\psi = -\sum_{\kappa=1}^{N} \frac{|\langle\psi_\kappa|\hat{H}_1|\psi_0\rangle|^2}{\varepsilon_\kappa} \qquad (10.46)$$

which is known as the sum-over-pseudostates expression. Equation 10.46 has the same form as the analogous expression that would arise from the discrete *eigenstates* of $\hat{H}_0$, but with definitely better convergence properties, reducing the infinite summation to a sum of a *finite* number of terms and avoiding the need of considering the contribution from the continuous part of the spectrum.

For *nonlinear* parameters, eventually occurring in $\tilde{\psi}_1$,[7] no specific expression can be given for the best energy $\tilde{E}_2$, which must be minimized with respect to these parameters by solving the *necessary* equations:

$$\frac{\partial\tilde{E}_2}{\partial c_\kappa} = 0, \quad \kappa = 1, 2, \ldots \qquad (10.47)$$

This can be done either analytically or numerically.

At this point, we can work out a few simple examples concerning the H atom in the external uniform *dipole* field (10.22), comparing the results of our approximate calculations of the polarizability α with the exact value given by (10.32). The results are collected in Table 10.1.

---

[7] Essentially, orbital exponents.

Table 10.1 Pseudostate approximations to α for the ground state of the H atom

φ	c	μ/ea_0	ε/E_h	α/a_0^3
$2p_z$	0.5	0.7449	0.375	2.96
$2p_z$	0.7970	0.9684	0.4191	4.48
$2p_z$ 1-term	1	1	0.5	4.0
$2p_z + 3p_z$ 2-term	1	$\begin{cases} 0.4082 \\ 0.9129 \end{cases}$	$\begin{cases} 1 \\ 0.4 \end{cases}$	$\begin{cases} 0.3333 \\ 4.1667 \end{cases}$
$2p_z + 3p_z + 4p_z$ 3-term	1	$\begin{cases} 0.8153 \\ 0.5572 \\ 0.0605 \end{cases}$	$\begin{cases} 0.3811 \\ 0.6166 \\ 1.7023 \end{cases}$	$\begin{cases} 3.4887 \\ 1.0070 \\ 0.0043 \end{cases}$

10.5.1 Single Pseudostate

Choose as a convenient variational function satisfying all symmetry prescriptions the normalized 2p STO:

$$\chi = \left(\frac{c^5}{\pi}\right)^{1/2} \exp(-cr)z, \quad z = r\cos\theta \tag{10.48}$$

We have

$$\tilde{\psi}_1 = \varphi = CF\chi, \quad \langle \psi_0|\chi\rangle = 0 \tag{10.49}$$

the order being carried by the *linear* parameter C, and omitted for brevity, while c is a *nonlinear* parameter.

Then, optimization against the single linear variation parameter (single-term approximation) gives

$$\tilde{E}_2 \text{ (best)} = -F^2 \frac{\mu^2}{\varepsilon} \tag{10.50}$$

where

$$\mu = \langle\chi|z|\psi_0\rangle = (\psi_0\chi|z) = \left(\frac{2\sqrt{c}}{c+1}\right)^5 \tag{10.51}$$

$$\varepsilon = \langle\chi|\hat{H}_0 - E_0|\chi\rangle = \frac{1}{2}(c^2 - c + 1) \tag{10.52}$$

In Equation 10.51, the first expression on the right-hand side is the transition moment in Dirac form and the second in charge density notation. These two integrals are evaluated in spherical coordinates in the Appendix.

$\tilde{E}_2$ can be further minimized with respect to the nonlinear parameter c entering both μ and ε. It is seen that *best c* is obtained as a solution of the cubic equation (Magnasco, 1978)

$$7c^3 - 9c^2 + 9c - 5 = 0 \tag{10.53}$$

which has the real root $c = 0.7970$.

We then have the cases exemplified in Table 10.1.

10.5.1.1   Eigenstate of $\hat{H}_0$

In this case we put $c = 1/2$ in expression (10.48). Evaluating the integrals by means of Equations 10.51 and 10.52 gives, in atomic units, the results of the *first* row of Table 10.1:

$$\mu = 0.7449 ea_0, \quad \varepsilon = \frac{3}{8} = 0.375 E_{\mathrm{h}}, \quad \alpha = 2.96 a_0^3 \tag{10.54}$$

This value of α is only 66% of the correct value 4.5. Including higher np_z eigenstates with $c = 1/n$ only improves this result a little. Including terms up to $n = 7$ gives $\alpha = 3.606$, the asymptotic value of $\alpha = 3.660$ being reached for $n = 30$. This is only 81.3% of the exact value, the remaining 18.7% coming from the contribution of the continuous part of the spectrum. These results show that the expansion in eigenstates of $\hat{H}_0$ is disappointingly poor, the correct value being obtained only through difficult calculations.

10.5.1.2   Single optimized pseudostate

The best $c = 0.7970$ is obtained in this case from the real root of Equation 10.53. We then obtain the results of the *second* row of Table 10.1:

$$\mu = 0.9684, \quad \varepsilon = 0.4191, \quad \alpha = 4.48 \tag{10.55}$$

α is now within 99.5% of the exact result. An enormous improvement over the eigenstate result is evident. The fact that the best orbital exponent c is sensibly larger than 0.5 suggests that the excited pseudostate must be rather contracted. It is then tempting to try with $c = 1$, giving a pseudostate having the same decay with r as the unperturbed ψ_0.

10.5.1.3 Single pseudostate with $c = 1$ (Single-term approximation)

The results are those given in the *third* row of Table 10.1:

$$\mu = 1, \quad \varepsilon = 0.5, \quad \alpha = 4 \tag{10.56}$$

This simple unoptimized pseudostate still gives 89% of the exact result, suggesting that a possible improvement would be to include further functions, suitably polarized along z, by means of the Ritz method, Equation 10.46.

10.5.2 N-term Approximation

The dipole pseudospectra of H(1s) for $N = 1$ through $N = 4$ are given in Table 10.2. The two-term approximation gives the exact result for the dipole polarizability α, the same being true for the three-term and the higher N-term ($N > 3$) approximations. In all such cases, the dipole polarizability of the atom is partitioned into an increasing number N of contributions arising from the different pseudostates:

$$\alpha = \sum_{i=1}^{N} \alpha_i \tag{10.57}$$

Table 10.2 Dipole pseudospectra of H(1s) for $N = 1$ through $N = 4$

i	α_i/a_0^3	ε_i/E_h	$\sum \alpha_i$
1	$4.000\,000 \times 10^0$	$5.000\,000 \times 10^{-1}$	4.0
1	$4.166\,667 \times 10^0$	$4.000\,000 \times 10^{-1}$	
2	$3.333\,333 \times 10^{-1}$	$1.000\,000 \times 10^0$	4.5
1	$3.488\,744 \times 10^0$	$3.810\,911 \times 10^{-1}$	
2	$9.680\,101 \times 10^{-1}$	$6.165\,762 \times 10^{-1}$	
3	$4.324\,577 \times 10^{-2}$	$1.702\,333 \times 10^0$	4.5
1	$3.144\,142 \times 10^0$	$3.764\,643 \times 10^{-1}$	
2	$1.091\,451 \times 10^0$	$5.171\,051 \times 10^{-1}$	
3	$2.564\,244 \times 10^{-1}$	$9.014\,629 \times 10^{-1}$	
4	$7.982\,236 \times 10^{-3}$	$2.604\,969 \times 10^0$	4.5

This increasingly fine subdivision of the exact polarizability value into different pseudostate contributions is of fundamental importance for the increasingly refined evaluation of the London dispersion coefficients for two H atoms interacting at long range, as we shall see in Chapter 11. The calculation of the two-term pseudostate approximation is fully described elsewhere (Magnasco, 2007) and will not be pursued further here. It can only be said that, in general, the N-term approximation will involve diagonalization of the $(N \times N)$ matrix $\mathbf{M}$ given by Equation 10.38, with eigenvalues giving the excitation energies ε_i and as eigenvectors the corresponding N-term pseudostates $\{\psi_i\}, i = 1, 2, \ldots, N$. For a given atom (or molecule), knowledge of the so-called N-term pseudospectrum $\{\alpha_i, \varepsilon_i\}, i = 1, 2, \ldots, N$, allows for the *direct* calculation of the dispersion coefficients of the interacting atoms (or molecules).

Before ending this section, we notice that Equation 10.27 gives the *exact* first-order solution of the RS dipole perturbation equation in un-normalized form. From this result we can obtain the single normalized pseudostate ψ *equivalent* to the exact ψ_1 as

$$\tilde{\psi}_1 = CF\psi, \quad \psi = N\left(z + \frac{r}{2}z\right)\psi_0 \qquad (10.58)$$

Easy calculation shows that

$$N = \sqrt{\frac{8}{43}}, \quad \mu = \frac{9}{2}\sqrt{\frac{2}{43}} = 0.9705, \quad \varepsilon = \frac{18}{43} = 0.4186, \quad \alpha = 4.5 \quad (10.59)$$

so that the single optimized pseudostate (10.55) is seen to overestimate the *true* ε by only $5 \times 10^{-4}E_h$ and underestimate the *true* μ by $0.0021ea_0$, giving a dipole polarizability which differs from the exact one by just $0.02\,a_0^3$. It must be admitted that the performance of this fully optimized *single* function is exceptionally good, but the linearly optimized two-term approximation does even better, giving the *exact* value for α (see Table 10.1).

10.6 QUANTUM THEORY OF MAGNETIC SUSCEPTIBILITIES

In this section we shall glance briefly at the use of RS perturbation techniques in treating magnetic susceptibilities of atoms and molecules.

An introductory presentation of the subject can be found in the book *Molecular Quantum Mechanics* (Atkins and Friedman, 2007), while a more specialistic presentation is given in *The Theory of the Electric and Magnetic Properties of Molecules* by Davies (1967). In the rest of this section we shall make use of both atomic and cgs/emu units.

The Hamiltonian for a particle of mass m and charge e in a *scalar* potential V is

$$\hat{H}_0 = \frac{\hat{p}^2}{2m} + V \tag{10.60}$$

where $\hat{p}$ is the operator for the linear momentum (impulse). In the presence of a magnetic perturbation due to a *vector* potential $\mathbf{A}$, $\hat{p}$ is replaced by

$$\hat{p} \Rightarrow \left(\hat{p} - \frac{e}{c}\mathbf{A}\right) \tag{10.61}$$

Hence, when acting on a function Ψ, the kinetic energy operator in the presence of $\mathbf{A}$ gives

$$
\begin{aligned}
\frac{\hat{p}^2}{2m}\Psi &= \frac{1}{2m}\left(\hat{p} - \frac{e}{c}\mathbf{A}\right)\cdot\left(\hat{p} - \frac{e}{c}\mathbf{A}\right)\Psi \\
&= \frac{1}{2m}\left(\hat{p}^2 - \frac{e}{c}\hat{\mathbf{p}}\cdot\mathbf{A} - \frac{e}{c}\mathbf{A}\cdot\hat{\mathbf{p}} + \frac{e^2}{c^2}\mathbf{A}^2\right)\Psi \\
&= -\frac{\hbar^2}{2m}\nabla^2\Psi + i\frac{e\hbar}{2mc}\nabla\cdot(\mathbf{A}\Psi) + i\frac{e\hbar}{2mc}\mathbf{A}\cdot\nabla\Psi + \frac{e^2}{2mc^2}\mathbf{A}^2\Psi \\
&= -\frac{\hbar^2}{2m}\nabla^2\Psi + i\frac{e\hbar}{2mc}(\mathbf{A}\cdot\nabla\Psi + \Psi\nabla\cdot\mathbf{A}) + i\frac{e\hbar}{2mc}\mathbf{A}\cdot\nabla\Psi + \frac{e^2}{2mc^2}\mathbf{A}^2\Psi \\
&= \left[-\frac{\hbar^2}{2m}\nabla^2 + i\frac{e\hbar}{mc}\mathbf{A}\cdot\nabla + i\frac{e\hbar}{2mc}(\nabla\cdot\mathbf{A}) + \frac{e^2}{2mc^2}\mathbf{A}^2\right]\Psi
\end{aligned}
\tag{10.62}
$$

Therefore, we obtain the correction terms to the unperturbed Hamiltonian (10.60) in the presence of a magnetic field:

$$\hat{H}_1 = i\frac{e\hbar}{mc}\mathbf{A}\cdot\nabla + i\frac{e\hbar}{2mc}(\nabla\cdot\mathbf{A}) \tag{10.63}$$

$$\hat{H}_2 = \frac{e^2}{2mc^2} A^2 \qquad (10.64)$$

with (10.63) being linear and (10.64) quadratic in **A**.

The vector potential **A** originates a magnetic field **H** at point **r**, given by the vector product

$$\mathbf{A} = \frac{1}{2}\mathbf{H} \times \mathbf{r} = \frac{1}{2}\begin{vmatrix} \mathbf{i} & \mathbf{j} & \mathbf{k} \\ H_x & H_y & H_z \\ x & y & z \end{vmatrix} \qquad (10.65)$$

where the components of the field are constant. Then, the last term in (10.63) is zero:

$$\nabla \cdot \mathbf{A} = \operatorname{div} \mathbf{A} = 0 \qquad (10.66)$$

Next, it can be easily shown that

$$(\mathbf{H} \times \mathbf{r}) \cdot \nabla = \mathbf{H} \cdot (\mathbf{r} \times \nabla) \qquad (10.67)$$

so that, taking into account spin, we get for $\hat{H}_1$:

$$\hat{H}_1 = -(\hat{\boldsymbol{\mu}}_L + g_e \beta_e \hat{\boldsymbol{\mu}}_S) \cdot \mathbf{H} \qquad (10.68)$$

where $g_e \approx 2$ is the intrinsic g-factor for the single electron (its correct value depends on considerations of quantum electrodynamics), β_e is the Bohr magneton, and $\hat{\boldsymbol{\mu}}_L$ and $\hat{\boldsymbol{\mu}}_S$ are the vector operators for the orbital and spin magnetic moments. It should be noted that the magnetic moments have a direction opposite to that of the vectors representing orbital and spin angular momenta.

For a magnetic field **H** *uniform* along z:

$$H_x = H_y = 0, \quad H_z = H, \quad \mathbf{H} = \mathbf{k}H \qquad (10.69)$$

where the field strength H should not be confused with the Hamiltonian symbol. Equation 10.65 then becomes

$$\mathbf{A} = \frac{1}{2}\mathbf{H} \times \mathbf{r} = \frac{1}{2}\begin{vmatrix} \mathbf{i} & \mathbf{j} & \mathbf{k} \\ 0 & 0 & H \\ x & y & z \end{vmatrix} = \frac{1}{2}H(-\mathbf{i}y + \mathbf{j}x) \qquad (10.70)$$

giving

$$A^2 = \mathbf{A} \cdot \mathbf{A} = \frac{1}{4} H^2 (x^2 + y^2) \qquad (10.71)$$

Hence

$$\hat{H}_2 = \frac{e^2}{2mc^2} A^2 = \frac{e^2}{8mc^2} H^2 (x^2 + y^2) \qquad (10.72)$$

For the energy of the system in the magnetic field we have the Taylor expansion in powers of the magnetic field (compare with the corresponding expansion in the electric field):

$$E(H) = E_0 + \left(\frac{\partial E}{\partial H} \right)_0 H + \frac{1}{2} \left(\frac{\partial^2 E}{\partial H^2} \right)_0 H^2 + \cdots$$

$$= E_0 - \mu_0 H - \frac{1}{2} \chi H^2 + \cdots \qquad (10.73)$$

where

$$\mu_0 = - \left(\frac{\partial E}{\partial H} \right)_0 \qquad (10.74)$$

is the permanent magnetic moment (due to orbital motion) and

$$\chi = - \left(\frac{\partial^2 E}{\partial H^2} \right)_0 \qquad (10.75)$$

is the magnetic susceptibility (which corresponds to the electric susceptibility).

10.6.1 Diamagnetic Susceptibilities

We consider first the ground states of spherically symmetric atoms ($m = 0$) and the singlet Σ ground states of diatomic molecules ($S = \Lambda = 0$).

(i) Atoms in spherical ground state
For atoms in a spherical ground state (such as hydrogen or rare gases) the first-order correction E_1 is zero (the nucleus is taken as the origin of

the vector potential **A**). In fact:

$$\hat{L}_z = -i\frac{\partial}{\partial\varphi} \quad \hat{L}_z Y_{\ell m}(\Omega) = m Y_{\ell m}(\Omega) \tag{10.76}$$

where

$$Y_{\ell m}(\Omega) \propto \exp(i\,m\varphi) P_\ell^m(\cos\theta) \tag{10.77}$$

is a spherical harmonic in complex form (Ω stands for the solid angle specified by θ and φ) and

$$m = 0, \pm 1, \pm 2, \ldots, \pm\ell \tag{10.78}$$

is the magnetic quantum number.

Then, using RS perturbation theory up to second order in the field H:

$$
\begin{aligned}
E_1 &= \langle\psi_0|\hat{H}_1|\psi_0\rangle = \langle\psi_0|-\hat{\boldsymbol{\mu}}_L \cdot \mathbf{H}|\psi_0\rangle \\
&= \langle\psi_0|\frac{e\hbar}{2mc}\hat{\mathbf{L}}\cdot\mathbf{H}|\psi_0\rangle = \frac{e\hbar}{2mc}H\langle\psi_0|\hat{L}_z|\psi_0\rangle = 0
\end{aligned}
\tag{10.79}
$$

since $m = 0$ for a spherical ground state;

$$
\begin{aligned}
E_2 &= \langle\psi_0|\hat{H}_2|\psi_0\rangle = \frac{e^2}{8mc^2}H^2\langle\psi_0|x^2+y^2|\psi_0\rangle \\
&= \frac{e^2}{12mc^2}H^2\langle r^2\rangle_{00}
\end{aligned}
\tag{10.80}
$$

and we obtain the Langevin contribution to the molar diamagnetic susceptibility:

$$\chi^L = -N_A \frac{e^2}{6mc^2}\langle r^2\rangle_{00} = -0.792 \times 10^{-6}\langle r^2\rangle_{00} < 0 \tag{10.81}$$

a negative contribution (when r is in au, the susceptibility is given in cgs/emu).

Table 10.3 gives the calculated diamagnetic susceptibilities for a few simple atoms. The value for the H atom is exact. For the two-electron atomic system, He, the hydrogenic approximation ($c = Z = 2$) gives

Table 10.3 Calculated diamagnetic susceptibilities for simple atoms

Atom	Method	$\langle r^2 \rangle_{00}/a_0^2$	$-\chi^L/10^{-6}$ cgs/emu
H	Exact	3	2.376
He	Hydrogenic	0.75	1.1880
	1-term SCF[a]	1.0535	1.6687
	2-term SCF[a]	1.1832	1.8743
	5-term SCF[a]	1.1847	1.8765
	Accurate[b]	1.1935	1.8905
Li	HF[c]		15.2
Be	HF[c]		14.1
Ne	HF[c]		7.4
Ar	HF[c]		20.9
Kr	HF[c]		33.0

[a] Clementi and Roetti, 1974.
[b] Pekeris, 1959.
[c] Strand and Bonham, 1964.

only 68 % of the accurate value (-1.8905×10^{-6} cgs/emu) obtained by Pekeris (1959). The single-term SCF (optimization of the orbital exponent c_0 of a minimum STO basis) improves the result to 88 %, while a five-term SCF (practically HF) gives over 99 % of the Pekeris result. Pretty good results are also obtained from the HF calculations (Strand and Bonham, 1964) on the heavier rare gases (7.4 instead of accurate 6.7 for Ne, 20.9 instead of 19.0 for Ar, 33.0 instead of 28.0 for Kr), showing a limited effect of electronic correlation on diamagnetic susceptibility.

For the high-frequency part, expanding the second-order energy in pseudostates $\{\psi_\kappa\}$, we have

$$
E_2^{\mathrm{hf}} = \langle \psi_0 | \hat{H}_1 | \psi_1 \rangle = - \sum_{\kappa(\neq 0)} \frac{\langle \psi_0 | \frac{e\hbar}{2mc} \hat{\mathbf{L}} \cdot \mathbf{H} | \psi_\kappa \rangle \langle \psi_\kappa | \frac{e\hbar}{2mc} \hat{\mathbf{L}} \cdot \mathbf{H} | \psi_0 \rangle}{E_\kappa - E_0}
$$

$$
= - \frac{e^2 \hbar^2}{4m^2 c^2} H^2 \sum_{\kappa(\neq 0)} \frac{|\langle \psi_0 | \hat{L}_z | \psi_\kappa \rangle|^2}{\varepsilon_\kappa}
$$

(10.82)

whence, referring to a mole:

$$
\chi^{\mathrm{hf}} = -\left(\frac{\partial^2 E}{\partial H^2} \right)_0 = N_A \frac{e^2 \hbar^2}{2m^2 c^2} \sum_{\kappa(\neq 0)} \frac{|\langle \psi_0 | \hat{L}_z | \psi_\kappa \rangle|^2}{\varepsilon_\kappa} > 0 \qquad (10.83)
$$

so that the high-frequency term (paramagnetic contribution to the diamagnetic susceptibility) is positive, of a sign opposite to that of the Langevin term. So, we have for the diamagnetic susceptibility

$$\chi^{d} = \chi^{L} + \chi^{hf} \qquad (10.84)$$

a molecular property independent of T. For atoms in spherical ground states, χ^{hf} vanishes, since all transition integrals are zero because of the orthogonality of the excited pseudostates ψ_{κ} to the ground state ψ_0:

$$\langle\psi_0|\hat{L}_z|\psi_{\kappa}\rangle = \langle 0|\hat{L}_z|\kappa\rangle = m\langle 0|\kappa\rangle = 0 \quad \text{for } \kappa \neq 0 \qquad (10.85)$$

(ii) Diatomic molecules in Σ singlet ground state
For molecules, $\chi^{hf} \neq 0$ and use is made of the average susceptibility:

$$\bar{\chi}^{d} = \frac{1}{3}\sum_{\alpha}\chi^{L}_{\alpha\alpha} + \frac{1}{3}\sum_{\alpha}\chi^{hf}_{\alpha\alpha} = \bar{\chi}^{L} + \bar{\chi}^{hf} \qquad (10.86)$$

where

$$\bar{\chi}^{L} = -N_A \frac{e^2}{6mc^2}\langle r^2\rangle_{00} < 0 \qquad (10.87)$$

$$\bar{\chi}^{hf} = N_A \frac{e^2\hbar^2}{2m^2c^2} \sum_{\kappa(\neq 0)} \frac{\left|\langle\psi_0|\hat{L}_z|\psi_{\kappa}\rangle\right|^2}{\varepsilon_{\kappa}} > 0 \qquad (10.88)$$

with $\varepsilon_{\kappa} > 0$ the excitation energy from the ground state $|0\rangle$ to the pseudostate $|\kappa\rangle$.

In Table 10.4 we give some values of diamagnetic susceptibilities for ground-state H_2 calculated with different wavefunctions (Tillieu, 1957a, 1957b). We see (i) that the high-frequency contribution is sensibly smaller than the low-fequency (Langevin) contribution, (ii) that the simple MO wavefunction exhibits an exceptionally good performance, comparing well with the accurate James–Coolidge wavefunction result of the last row, (iii) that the HL (purely covalent) wavefunction shows a reasonable behaviour, while (iv) the Weinbaum (HL plus ionic) wavefunction gives results that are definitely too high either for χ^{L} or χ^{hf}.

Table 10.4 Calculated[a] diamagnetic susceptibilities for ground-state H_2 (units 10^{-6} cgs/emu)

Wavefunction	χ^L	χ^{hf}	χ^d
MO	−4.03	0.09	−3.94
HL	−4.22	0.11	−4.11
Weinbaum	−5.55	0.26	−5.29
James–Coolidge	−3.92	0.07	−3.85

[a] Tillieu,1957a, 1957b.

In Table 10.5 we give some calculated values of diamagnetic susceptibilities for the singlet Σ ground state of a few simple diatomics. Some experimental values are given in parentheses. The agreement with experiment can be judged as moderately satisfactory, even if the high-frequency contributions calculated for the isoelectronic molecules N_2 and CO are definitely too low.

10.6.2 Paramagnetic Susceptibilities

We now pass to consider briefly electronic states of atoms, ions and molecules in nonsinglet spin states $(S \neq 0)$, originating T-dependent paramagnetic susceptibilities in gases. In these cases, Curie's law holds:

$$\chi^p = \frac{C_m}{T} \qquad (10.89)$$

Table 10.5 Calculated diamagnetic susceptibilities for the ground state of simple diatomics (units 10^{-6} cgs/emu)

Molecule	χ^{La-c}	χ^{hfc}
Li_2	−83.027	50.53
N_2	−40.800(−43.6)	23.47(30.3)
F_2	−64.800	46.15
LiH	−20.529	11.38(12.5)
FH	−9.588(−9.2)	0.937(0.61)
LiF		16.10(15.3)
CO		21.94(28.2)

[a] Stevens, Pitzer, and Lipscomb, 1963.
[b] Stevens and Lipscomb, 1964a, 1964b, 1964c.
[c] Karplus and Kolker, 1961, 1963.

where

$$C_m = N_A \frac{\mu^2}{3k} \qquad (10.90)$$

is the Curie constant (referred to a mole).

Adding the diamagnetic susceptibility χ^d, we have for the *molar* magnetic susceptibility

$$\chi^m = \chi^d + \chi^p = -A + \frac{C}{T} \qquad (10.91)$$

a result similar to that existing for electric polarizabilities.

For paramagnetic systems, where often electronic spin plays a fundamental role in determining susceptibilities, the first term in (10.91) is usually negligible with respect to χ^p (hence Curie's law).

A generalization of (10.89) is given by the law by Curie–Weiss:

$$\chi^m = \frac{C_m}{T - \Theta} \Rightarrow \frac{1}{\chi^m} = -\frac{\Theta}{C_m} + \frac{1}{C_m} T \qquad (10.92)$$

where Θ ($< T$) is called the Curie temperature. Θ is a quantity characteristic of the different substances, its value marking the difference between paramagnetic ($\Theta < 0$) and ferromagnetic ($\Theta > 0$) systems. Equation 10.92 shows that the reciprocal of the magnetic susceptibility is *linear* in the temperature T and can be used for the experimental determination of the Curie constant (slope) and the Curie temperature (intercept).

(i) Paramagnetic susceptibilities

To investigate on the value assumed by the elementary magnetic moment μ in the case of light atoms and ions, we may resort to the Russell–Saunders LS-coupling scheme in the so-called vector model (Magnasco, 2007). In this scheme, the spin vectors **L** and **S** are coupled to a resultant vector **J** having a component J_z along the direction of the magnetic field. The associated 'good' quantum numbers J and M_J take the values

$$\begin{cases} J = L + S, L + S - 1, \ldots, |L - S| \\ M_J = -J, -(J-1), \ldots, (J-1), J \end{cases} \qquad (10.93)$$

QUANTUM THEORY OF MAGNETIC SUSCEPTIBILITIES

The modulus (absolute value or magnitude) of the magnetic moment of the atom in state J will be

$$\mu_J = -g_e \frac{e\hbar}{2mc} \sqrt{J(J+1)} = -g_e \beta_e \sqrt{J(J+1)} \qquad (10.94)$$

where g_e is now the Landé g-factor:

$$g_e = 1 + \frac{J(J+1) + S(S+1) - L(L+1)}{2J(J+1)} \qquad (10.95)$$

For a single s-electron, $L = 0$, $S = \frac{1}{2}$, $J = S = \frac{1}{2}$, and therefore $g_e = 2$. The average molar magnetic moment for an atom in the quantum state J for a *small* field will hence be

$$\langle \mu_J \rangle = N_A \frac{\mu^2}{3kT} H = N_A \frac{g_e^2 \beta_e^2}{3kT} J(J+1) H \qquad (10.96)$$

giving for the T-dependent molar paramagnetic susceptibility χ^p

$$\chi^p = \frac{\langle \mu_J \rangle}{H} = N_A \frac{g_e^2 \beta_e^2}{3kT} J(J+1) \qquad (10.97)$$

As already said, in most cases paramagnetism is given by spin only ($g_e = 2$), so that

$$\chi^s = \frac{\langle \mu_S \rangle}{H} = 4 N_A \frac{\beta_e^2}{3kT} S(S+1) \qquad (10.98)$$

The effective magnetic moment for an atom (or ion) in state J in units of the Bohr magneton β_e will be

$$\frac{\mu_{\text{eff}}}{\beta_e} = g_e \sqrt{J(J+1)} \qquad (10.99)$$

Spectroscopically, there are three cases according to the value of the *multiplet width* $\Delta = h\nu$ with respect to the temperature T.

(a) $\Delta \ll kT$ (*narrow* multiplets)
We have

$$\frac{\mu_{\text{eff}}}{\beta_e} = [4J(J+1)]^{1/2} \qquad (10.100)$$

This is the case of the ions of the metals of the first transition group, and of the triplet ground state of O_2. If L and S are quantized independently (like in Fe^{2+}), then

$$\frac{\mu_{\text{eff}}}{\beta_e} = [4S(S+1) + L(L+1)]^{1/2} \qquad (10.101)$$

(b) $\Delta \gg kT$ (*wide* multiplets)
Now, almost all particles are in the state of lowest energy J:

$$\frac{\mu_{\text{eff}}}{\beta_e} = g_e[J(J+1)]^{1/2}, \quad g_e = \frac{3}{2} + \frac{S(S+1)-L(L+1)}{2J(J+1)}$$

$$(10.102)$$

the case of the rare-earth ions.

(c) $\Delta \approx kT$ (at room temperature, $T = 293\,\text{K}$)
This is the most difficult case (for example, NO, Sm^{3+} and Eu^{3+}), and it has been treated in detail by Van Vleck (1932).

We now discuss briefly a few interesting cases of paramagnetism in atoms, molecules and ions. In comparing with experiment, it is convenient to put

$$N_B = N_A \frac{e\hbar}{2mc} = N_A\beta_e = 5.585 \times 10^3 \frac{\text{erg}}{\text{gauss mol}} \qquad (10.103)$$

1 mol of Bohr magnetons, and an effective magnetic moment μ_{eff} defined through

$$\chi_m = N_A \frac{g_e^2 \beta_e^2}{3kT} J(J+1) = \frac{g_e^2 N_B^2}{3RT} J(J+1) = \frac{(N_B\mu_{\text{eff}})^2}{3RT} \qquad (10.104)$$

where R is the gas constant, so that

$$\mu_{\text{eff}}^2 = \frac{3R}{N_B^2}(\chi_m T) = [g_e \sqrt{J(J+1)}]^2 \qquad (10.105)$$
$$\text{experiment} \qquad \text{theory}$$

$$\mu_{\text{eff}} = \left(\frac{3R}{N_B^2}\right)^{1/2} \sqrt{\chi_m T} = 2.828 \sqrt{\chi_m T} \qquad (10.106)$$

(ii) Atoms in S states

For the H atom in its ^2S ground state, $L = 0$ and paramagnetism is due to spin only:

$$\begin{cases} L = 0, \quad J = S = \dfrac{1}{2}, \quad g_e = 2 \\[2mm] \mu_S = -g_e \beta_e \sqrt{S(S+1)} = -\beta_e \sqrt{3} \\[2mm] \mu_{\text{eff}} = \dfrac{|\mu_S|}{\beta_e} = \sqrt{3} \approx 1.73 \end{cases} \qquad (10.107)$$

It has been observed that all complete atomic shells (with an electron configuration similar to that of rare gases) have $J = 0$, so that the only magnetically active electrons are those of the valence (or optical) shell. For instance, for vapours of atomic Na, we have only a 3s-valence electron, so that the situation is entirely similar to that of the ground-state H atom, giving

$$\mu_S = -\sqrt{3} N_B \Rightarrow \mu_{\text{eff}} = \frac{|\mu_S|}{N_B} = \sqrt{3} \approx 1.73 \qquad (10.108)$$

Experimental measurements of $\chi_m T$ in the temperature range 900–1000 K give 0.38, so that

$$\mu_{\text{eff}} = 2.828 \sqrt{\chi_m T} \approx 1.74 \qquad (10.109)$$

which is in almost perfect agreement with the theoretical value of $\sqrt{3}$.

(iii) **The $^3\Sigma_g^-$ ground state of the O_2 molecule**
We recall that

$$1\, cm^{-1} = \frac{2625.5}{219.47 \times 10^3}\, kJ = \frac{315.78 \times 10^3}{219.47 \times 10^3}\, K = 1.439\, K \quad (10.110)$$

Molecular Σ states are practically devoided of multiplet structure, although experimentally they do have a small fine structure of the order of $1\, cm^{-1}$ or less. Hence, χ^P can certainly be calculated under the assumption that the multiplet structure is small compared with kT (case a). At room temperature ($T = 293\, K$):

$$\frac{\Delta E}{kT} = \frac{1.44}{293} \approx 4.9 \times 10^{-3} \quad (10.111)$$

and the assumptions of Curie's law are valid. For the triplet $^3\Sigma_g^-$ ground state of O_2, $\Lambda = 0$ and $S = 1$, so that

$$\chi^P = \frac{N_A \beta_e^2}{3kT} \mu_{eff}^2 = \frac{N_A \beta_e^2}{3kT}\left[4S(S+1) + \Lambda^2\right] = \frac{N_A \beta_e^2}{3k}\frac{8}{T} \quad (10.112)$$

Therefore:

$$\frac{N_A \beta_e^2}{3k} \approx 0.125 \Rightarrow 8\frac{N_A \beta_e^2}{3k} \approx 1 \Rightarrow \chi^P \approx \frac{1}{T} \quad (10.113)$$

At $T = 293\, K$, the theory gives

$$\chi^P(^3\Sigma) \approx 3.41 \times 10^{-3}\, cgs/emu \quad (10.114)$$

which is almost in perfect agreement with the average experimental value of 3.408×10^{-3}. Original work by Curie himself shows that Curie's law for $O_2(^3\Sigma)$ is very well satisfied over the temperature range 290–720 K.

(iv) **A free ^{3}F ion without LS-coupling in a magnetic field H**
The magnetic moments operators corresponding to the quantum states characterized by *uncoupled* L and S are

$$\hat{\boldsymbol{\mu}}_L = -\beta_e \hat{\mathbf{L}}, \quad \hat{\boldsymbol{\mu}}_S = -2\beta_e \hat{\mathbf{S}} \tag{10.115}$$

Then, the potential energy of the magnetic dipole in the uniform magnetic field $\mathbf{H} = \mathbf{k}\,H$ is

$$\hat{H}_1 = -(\hat{\boldsymbol{\mu}}_L + \hat{\boldsymbol{\mu}}_S) \cdot \mathbf{H} = \beta_e H(\hat{L}_z + 2\hat{S}_z) \tag{10.116}$$

with

$$\hat{L}_z\psi = M_L\psi, \quad \hat{S}_z\psi = M_S\psi, \quad \psi = \psi(M_L, M_S) \tag{10.117}$$

so that the Zeeman energy splitting of the $(2L+1)(2S+1)$-sublevels in presence of the field $\mathbf{F}$ will be $(-L \leq M_L \leq L, \ -S \leq M_S \leq S)$

$$\Delta E(M_L, M_S) = \beta_e H(M_L + 2M_S) \tag{10.118}$$

There are altogether $(2L+1)(2S+1) = 7 \times 3 = 21$ energy levels, seven of which are still *degenerate* even in presence of the field (degeneracies 2, 2, 3, 3, 3, 2, 2), as shown schematically in Figure 10.2.
(v) **An LS-coupled ^{3}F ion in a magnetic field H**
In this case:

$$\begin{cases} L = 3, \ S = 1, \ J = 4, 3, 2 \\[2mm] g_e = \dfrac{3}{2} - \dfrac{5}{J(J+1)} \\[2mm] J = 4, \ g_e = \dfrac{5}{4}, \quad J = 3, \ g_e = \dfrac{13}{12}, \quad J = 2, \ g_e = \dfrac{2}{3} \end{cases} \tag{10.119}$$

There are $9 + 7 + 5 = 21$ levels altogether, as before, but now in the presence of a field $\mathbf{H}$ any degeneracy is removed.

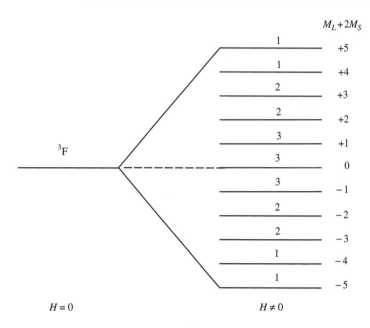

Figure 10.2 Zeeman splitting for a free ^3F ion in a magnetic field H (the residual degeneracy of each level is shown)

The operator describing spin-orbital LS-coupling is

$$\hat{H}^{SO} = g_e\beta_e^2\frac{Z}{r^3}\hat{L}\cdot\hat{S} = \xi(r)\hat{L}\cdot\hat{S} \qquad (10.120)$$

where Z is the nuclear charge of the ion. But:

$$\hat{L}\cdot\hat{S} = \frac{1}{2}(J^2-L^2-S^2) \qquad (10.121)$$

Using first-order perturbation theory, we take the expectation value of $\hat{H}^{SO}$ over the ground-state wavefunction and the relative energy of levels of given J will be

$$E_J = A\frac{1}{2}[J(J+1)-L(L+1)-S(S+1)] \qquad (10.122)$$

where A is a constant, characteristic of the ion. We then get the relative energies of the three LS-coupled states occurring for $L = 3$ (state F)

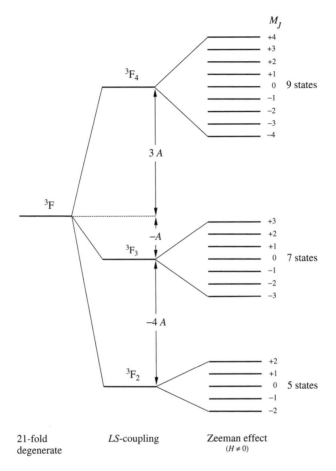

Figure 10.3 Resolution of the multiplet structure of the spin-orbit coupled ^3F_J state in a magnetic field H (all levels now have different energies)

and $S = 1$ (triplet):

$$\begin{cases} J = 4, & E_4 = 3A \\ J = 3, & E_4 = -A \\ J = 2, & E_4 = -4A \end{cases} \tag{10.123}$$

(vi) **The Zeeman effect for the ^3F_J ion in a magnetic field H**
The magnetic moment operator corresponding to the quantum state J resulting from LS-coupling is

$$\hat{\boldsymbol{\mu}}_J = -g_e\beta_e\hat{\mathbf{J}}, \quad g_e = \frac{3}{2} + \frac{S(S+1)-L(L+1)}{2J(J+1)} \tag{10.124}$$

Then, the potential energy of the magnetic dipole in the uniform magnetic field $\mathbf{H} = \mathbf{k}\,H$ is

$$\begin{cases} \hat{H}_1 = -\hat{\boldsymbol{\mu}}_J \cdot \mathbf{H} = g_e\beta_e H\hat{J}_z \\ \hat{J}_z\psi = M_J\psi \quad (-J \le M_J \le J) \end{cases} \tag{10.125}$$

so that the energy of the $(2J+1)$-sublevels in presence of a field H will be

$$E_{M_J} = g_e\beta_e HM_J \tag{10.126}$$

The splitting of the Zeeman levels is linear in the strength H of the magnetic field, and is shown schematically in Figure 10.3.

APPENDIX: EVALUATION OF μ AND ε

For the STO (10.48), the required integrals are easily calculated in spherical coordinates.

$$\mu = \langle 2p_z|z|\psi_0\rangle = (\psi_0 2p_z|z) = \frac{\sqrt{c^5}}{\pi}2\pi\int_{-1}^{1}dx\,x^2\int_{0}^{\infty}dr\,r^4\exp[-(c+1)r]$$

$$= \frac{4}{3}\sqrt{c^5}\frac{4\times 3\times 2}{(c+1)^5} = \left(\frac{2\sqrt{c}}{c+1}\right)^5 \tag{10.127}$$

Using relations (10.29), we obtain

$$\begin{cases} \dfrac{d}{dr}[\exp(-cr)r] = \exp(-cr)(1-cr) \\[2mm] \dfrac{d^2}{dr^2}[\exp(-cr)r] = \exp(-cr)(-2c+c^2r) \\[2mm] \nabla_r^2[\exp(-cr)r] = \exp(-cr)\left(\dfrac{2}{r}-4c+c^2r\right) \end{cases} \tag{10.128}$$

so that

$$\hat{H}_0[\exp(-cr)r\cos\theta] = \exp(-cr)\left[(2c-1)-\frac{c^2}{2}r\right]\cos\theta \qquad (10.129)$$

We then obtain for the integral:

$$\langle 2p_z|\hat{H}_0|2p_z\rangle = \frac{c^5}{\pi}2\pi\int_{-1}^{1}dx\,x^2\int_{0}^{\infty}dr\,r^2\left[(2c-1)r-\frac{c^2}{2}r^2\right]\exp(-2cr)$$

$$= \frac{4}{3}c^5\left[(2c-1)\frac{3\times2}{(2c)^4}-\frac{c^2}{2}\frac{4\times3\times2}{(2c)^5}\right] = \frac{c^2}{2}-\frac{c}{2}$$

$$(10.130)$$

giving for the excitation energy

$$\varepsilon = \langle 2p_z|\hat{H}_0-E_0|2p_z\rangle = \frac{1}{2}(c^2-c+1) \qquad (10.131)$$

11

Atomic and Molecular Interactions

In this chapter we shall present first (Section 11.1) the elementary RS perturbation theory for the interaction between two ground-state H atoms, taking the interatomic potential V as a first-order perturbation. The interacting atoms are assumed to be sufficiently far apart so that, in the first approximation, we can ignore the effect of exchanging identical electrons between different atoms, as required by Pauli's antisymmetry principle. As a consequence, the two-electron wavefunctions will be taken in the form of simple orbital products for the two electrons, insisting on the fact that electron 1 belongs to atom A and electron 2 to atom B, giving what is called the Coulombic interatomic energy, which will be considered up to second order in V. The possibility of including electron exchange in the first-order theory is examined in some detail elsewhere (Magnasco, 2007), where it is seen that it is equivalent to the HL theory of Chapter 9. The expansion of the interatomic potential V into inverse powers of the internuclear distance R, giving the so-called multipole expansion of the potential, is then examined in Section 11.2, with particular emphasis on the calculation of the leading term of the London attraction between two H atoms. These considerations are then extended to molecules (Section 11.3), while a short discussion on the nature of the VdW and hydrogen bonds concludes the chapter.

Methods of Molecular Quantum Mechanics: An Introduction to Electronic Molecular Structure
Valerio Magnasco
© 2009 John Wiley & Sons, Ltd

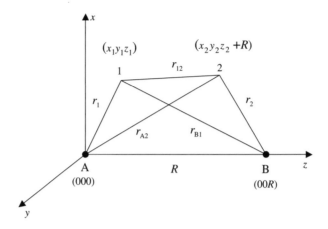

Figure 11.1   Interatomic reference system for the H–H interaction

11.1   THE H–H NONEXPANDED INTERACTIONS UP TO SECOND ORDER

With reference to Figure 11.1, we take as unperturbed Hamiltonian $\hat{H}_0$ the sum of the two Hamiltonians for the *separate* H atoms having unperturbed wavefunctions $a_0(\mathbf{r}_1)$ and $b_0(\mathbf{r}_2)$ and energies E_0^A and E_0^B:

$$\hat{H}_0 = \hat{H}_0^A + \hat{H}_0^B, \quad \psi_0 = a_0(\mathbf{r}_1)b_0(\mathbf{r}_2), \quad E_0 = E_0^A + E_0^B \tag{11.1}$$

The perturbation $\hat{H}_1$ is here the *interatomic potential V*:

$$V = \hat{H} - \hat{H}_0 = -\frac{1}{r_{B1}} - \frac{1}{r_{A2}} + \frac{1}{r_{12}} + \frac{1}{R} \tag{11.2}$$

where R is the internuclear distance measured along the z-axis. We shall refer to (11.2) as the *nonexpanded* interatomic potential.

If we introduce a set of *excited pseudostates* $\{a_i\}$ on A and $\{b_j\}$ on B $(i, j = 1, 2, \ldots, N)$, normalized and orthogonal to the respective unperturbed functions:

$$\begin{cases} \langle a_i|a_i \rangle = \langle b_j|b_j \rangle = 1 \\ \langle a_i|a_0 \rangle = \langle a_0|a_i \rangle = \langle b_j|b_0 \rangle = \langle b_0|b_j \rangle = 0 \end{cases} \tag{11.3}$$

then the RS energy corrections up to second order in V are as follows:

$$E_0 = \langle\psi_0|\hat{H}_0|\psi_0\rangle = \langle a_0 b_0|\hat{H}_0^A + \hat{H}_0^B|a_0 b_0\rangle = E_0^A + E_0^B \qquad (11.4)$$

the unperturbed energy, the energy pertaining to the separate atoms;

$$E_1^{cb} = \langle\psi_0|\hat{H}_1|\psi_0\rangle = \langle a_0 b_0|V|a_0 b_0\rangle$$
$$= (a_0^2| - r_{B1}^{-1}) + (b_0^2| - r_{A2}^{-1}) + (a_0^2|b_0^2) + \frac{1}{R} = E_1^{es} \qquad (11.5)$$

the nonexpanded first-order correction, the semiclassical Coulomb interaction of the HL theory, said the *electrostatic* energy; and

$$E_2^{cb} = -\sum_i \frac{|\langle a_i b_0|V|a_0 b_0\rangle|^2}{\varepsilon_i} - \sum_j \frac{|\langle a_0 b_j|V|a_0 b_0\rangle|^2}{\varepsilon_j}$$
$$- \sum_i \sum_j \frac{|\langle a_i b_j|V|a_0 b_0\rangle|^2}{\varepsilon_i + \varepsilon_j} \qquad (11.6)$$

the second-order Coulombic energy describing deviations from the rigid spherical atoms. The first two terms represent the *induction* (distortion or polarization) energy involving *monoexcitations* (top row of Figure 11.2),

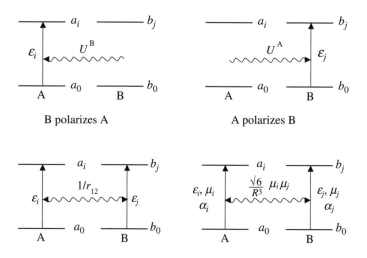

Mutual polarization of A ⇔ B

Figure 11.2 Second-order interactions between ground-state H atoms. Top row: B polarizes A (left) and A polarizes B (right). Bottom row: nonexpanded dispersion (left) and leading term of the expanded dispersion (right)

ATOMIC AND MOLECULAR INTERACTIONS

one on A (B polarizes A) and the other on B (A polarizes B), the last term the *dispersion* (interatomic electron correlation) energy involving simultaneous *biexcitations* (bottom row of Figure 11.2), one on A and the other on B.

Introducing the explicit form of V, Equation 11.2, we obtain for the transition integrals in (11.6)

$$\langle a_i b_0 | V | a_0 b_0 \rangle = \left\langle a_i b_0 \left| -\frac{1}{r_{B1}} - \frac{1}{r_{A2}} + \frac{1}{r_{12}} + \frac{1}{R} \right| a_0 b_0 \right\rangle$$

$$= \left\langle a_i \left| -\frac{1}{r_{B1}} + \int d\mathbf{r}_2 \frac{[b_0(\mathbf{r}_2)]^2}{r_{12}} \right| a_0 \right\rangle = (a_0 a_i | U^B) \qquad (11.7)$$

where

$$U^B(\mathbf{r}_1) = -\frac{1}{r_{B1}} + \int d\mathbf{r}_2 \frac{[b_0(\mathbf{r}_2)]^2}{r_{12}} = -\frac{\exp(-2r_{B1})}{r_{B1}}(1 + r_{B1}) \qquad (11.8)$$

is the *molecular electrostatic potential* (MEP) at $\mathbf{r}_1$ on A due to B (nucleus and undistorted electron of B).

Similarly:

$$\langle a_0 b_j | V | a_0 b_0 \rangle = \left\langle a_0 b_j \left| -\frac{1}{r_{B1}} - \frac{1}{r_{A2}} + \frac{1}{r_{12}} + \frac{1}{R} \right| a_0 b_0 \right\rangle$$

$$= \left\langle b_j \left| -\frac{1}{r_{A2}} + \int d\mathbf{r}_1 \frac{[a_0(\mathbf{r}_1)]^2}{r_{12}} \right| b_0 \right\rangle = (b_0 b_j | U^A) \qquad (11.9)$$

with

$$U^A(\mathbf{r}_2) = -\frac{1}{r_{A2}} + \int d\mathbf{r}_1 \frac{[a_0(\mathbf{r}_1)]^2}{r_{12}} = -\frac{\exp(-2r_{A2})}{r_{A2}}(1 + r_{A2}) \qquad (11.10)$$

the MEP at $\mathbf{r}_2$ on B due to atom A in its undistorted ground state.

For the transition integral involving the biexcitations we obtain

$$\langle a_i b_j | V | a_0 b_0 \rangle = \left\langle a_i b_j \left| -\frac{1}{r_{B1}} - \frac{1}{r_{A2}} + \frac{1}{r_{12}} + \frac{1}{R} \right| a_0 b_0 \right\rangle$$
$$= \left\langle a_i b_j \left| \frac{1}{r_{12}} \right| a_0 b_0 \right\rangle = (a_0 a_i | b_0 b_j) \tag{11.11}$$

the interaction between the two *transition* densities $[a_0(\mathbf{r}_1) a_i^*(\mathbf{r}_1)]$ on A and $[b_0(\mathbf{r}_2) b_j^*(\mathbf{r}_2)]$ on B.

The first few orders of *nonexpanded* RS perturbation theory for the H–H interaction are then

$$E_1^{cb} = E_1^{es} = \frac{\exp(-2R)}{R} \left(1 + \frac{5}{8}R - \frac{3}{4}R^2 - \frac{1}{6}R^3 \right) \tag{11.12}$$

the first-order electrostatic energy, with its explicit dependence on the internuclear distance R, showing the *charge-overlap* nature of the interaction between neutral H atoms.

For the second-order energy:

$$\tilde{E}_2 = \tilde{E}_2^{ind,A} + \tilde{E}_2^{ind,B} + \tilde{E}_2^{disp} \tag{11.13}$$

where

$$\tilde{E}_2^{ind,A} = -\sum_i \frac{|(a_0 a_i | U^B)|^2}{\varepsilon_i} \tag{11.14}$$

is the polarization energy of A by B (B distorts A from its spherical symmetry),

$$\tilde{E}_2^{ind,B} = -\sum_j \frac{|(b_0 b_j | U^A)|^2}{\varepsilon_j} \tag{11.15}$$

is the polarization energy of B by A (A distorts B from its spherical symmetry) and

$$\tilde{E}_2^{disp} = -\sum_i \sum_j \frac{|(a_0 a_i | b_0 b_j)|^2}{\varepsilon_i + \varepsilon_j} \tag{11.16}$$

is the *nonexpanded* dispersion energy due to the mutual polarization (distortion) of A and B. The tilde on E_2 and its components means that we are using a variational approximation to the second-order energies.

11.2 THE H–H EXPANDED INTERACTIONS UP TO SECOND ORDER

We now expand the interatomic potential V at long range $(R \gg r_1, r_2)$ in a power series in R^{-n} up to R^{-3}:

$$
\begin{aligned}
\frac{1}{r_{B1}} &= [x_1^2 + y_1^2 + (z_1 - R)^2]^{-1/2} = (R^2 - 2z_1R + r_1^2)^{-1/2} \\
&= \frac{1}{R}\left(1 - 2\frac{z_1}{R} + \frac{r_1^2}{R^2}\right)^{-1/2} \\
&\approx \frac{1}{R}\left[1 + \frac{z_1}{R} - \frac{r_1^2}{2R^2} + \frac{3}{8}\left(-2\frac{z_1}{R}\right)^2 + \cdots\right] \\
&= \frac{1}{R} + \frac{z_1}{R^2} + \frac{3z_1^2 - r_1^2}{2R^3} + O(R^{-4})
\end{aligned}
\tag{11.17}
$$

where use was made of the Taylor expansion for $x = $ small:

$$
(1 + x)^{-1/2} \approx 1 - \frac{1}{2}x + \frac{3}{8}x^2 + \cdots
\tag{11.18}
$$

$$
\begin{aligned}
\frac{1}{r_{A2}} &= [x_2^2 + y_2^2 + (z_2 + R)^2]^{-1/2} = (R^2 + 2z_2R + r_2^2)^{-1/2} \\
&= \frac{1}{R}\left(1 + 2\frac{z_2}{R} + \frac{r_2^2}{R^2}\right)^{-1/2} \\
&\approx \frac{1}{R}\left[1 - \frac{z_2}{R} - \frac{r_2^2}{2R^2} + \frac{3}{8}\left(2\frac{z_2}{R}\right)^2 + \cdots\right] \\
&= \frac{1}{R} - \frac{z_2}{R^2} + \frac{3z_2^2 - r_2^2}{2R^3} + O(R^{-4})
\end{aligned}
\tag{11.19}
$$

For the two-electron repulsion, we have

$$
\begin{aligned}
\frac{1}{r_{12}} &= [(x_1 - x_2)^2 + (y_2 - y_2)^2 + (z_1 - z_2 - R)^2]^{-1/2} \\
&= (R^2 - 2z_1 R + 2z_2 R + r_1^2 + r_2^2 - 2x_1 x_2 - 2y_1 y_2 - 2z_1 z_2)^{-1/2} \\
&= \frac{1}{R}\left[1 - 2\frac{z_1}{R} + 2\frac{z_2}{R} + \frac{r_1^2}{R^2} + \frac{r_2^2}{R^2} - \frac{2}{R^2}(x_1 x_2 + y_1 y_2 + z_1 z_2) \right]^{-1/2}
\end{aligned}
$$

$$(11.20)$$

so that, expanding according to Taylor:

$$
\begin{aligned}
\frac{1}{r_{12}} &\approx \frac{1}{R}\left[1 + \frac{z_1}{R} - \frac{z_2}{R} - \frac{r_1^2}{2R^2} - \frac{r_2^2}{2R^2} + \frac{1}{R^2}(x_1 x_2 + y_1 y_2 + z_1 z_2) \right. \\
&\left. \quad + \frac{3}{8}\left(-2\frac{z_1}{R}\right)^2 + \frac{3}{8}\left(2\frac{z_2}{R}\right)^2 + \frac{3}{8}\left(-8\frac{z_1 z_2}{R^2}\right) + \cdots \right] \\
&= \frac{1}{R} + \frac{z_1 - z_2}{R^2} + \frac{3z_1^2 - r_1^2}{2R^3} + \frac{3z_2^2 - r_2^2}{2R^3} + \frac{x_1 x_2 + y_1 y_2 - 2z_1 z_2}{R^3} + O(R^{-4})
\end{aligned}
$$

$$(11.21)$$

Adding all terms altogether with the appropriate signs, many terms do cancel, finally giving:

$$
V \approx \frac{1}{R^3}(x_1 x_2 + y_1 y_2 - 2z_1 z_2) + O(R^{-4}) \tag{11.22}
$$

which is the leading term, the dipole-dipole interaction, of the *expanded* form of the interatomic potential V for neutral H atoms. It corresponds to the classical electrostatic interaction of two point-like dipoles[1] located at the two nuclei of A and B (Coulson, 1958). Expansion (11.22) is the first term of what is known as the *multipole expansion* of the interatomic potential in long range.

---

[1] In atomic units.

Then, with such an expanded V, it is easily seen that

$$E_1^{cb} = E_1^{es} = 0 \tag{11.23}$$

$$\tilde{E}_2^{ind,A} = \tilde{E}_2^{ind,B} = 0 \tag{11.24}$$

so that the only surviving term at long range is the London *dispersion* attraction:

$$
\begin{aligned}
\tilde{E}_2^{disp} &= -\frac{1}{R^6}\sum_i\sum_j\frac{|\langle a_ib_j|x_1x_2 + y_1y_2 - 2z_1z_2|a_0b_0\rangle|^2}{\varepsilon_i + \varepsilon_j} \\
&= -\frac{6}{R^6}\sum_i\sum_j\frac{|(a_0a_i|z_1)|^2|(b_0b_j|z_2)|^2}{\varepsilon_i + \varepsilon_j}
\end{aligned}
\tag{11.25}
$$

where we have taken into account the spherical symmetry of atoms A and B, giving

$$
\begin{cases}
(a_0a_i|x_1) = (a_0a_i|y_1) = (a_0a_i|z_1) & \text{on A} \\
(b_0b_j|x_2) = (b_0b_j|y_2) = (b_0b_j|z_2) & \text{on B}
\end{cases}
\tag{11.26}
$$

Since

$$\alpha_i^A = 2\frac{|(a_0a_i|z_1)|^2}{\varepsilon_i} = 2\frac{\mu_i^2}{\varepsilon_i}, \qquad \alpha^A = \sum_i\alpha_i^A \tag{11.27}$$

is the ith pseudostate contribution to α^A, the dipole polarizability of atom A, and

$$\alpha_j^B = 2\frac{|(b_0b_j|z_2)|^2}{\varepsilon_j} = 2\frac{\mu_j^2}{\varepsilon_j}, \qquad \alpha^B = \sum_j\alpha_j^B \tag{11.28}$$

is the jth pseudostate contribution to α^B, the dipole polarizability of atom B, the formula for the leading term of the expanded dispersion can be written in the so-called London form:

$$
\begin{aligned}
\tilde{E}_2^{disp} &= -\frac{6}{R^6}\sum_i\sum_j\frac{\mu_i^2\mu_j^2}{\varepsilon_i + \varepsilon_j} = -\frac{6}{R^6}\frac{1}{4}\sum_i\sum_j\left(\frac{2\mu_i^2}{\varepsilon_i}\right)\left(\frac{2\mu_j^2}{\varepsilon_j}\right)\frac{\varepsilon_i\varepsilon_j}{\varepsilon_i + \varepsilon_j} \\
&= -\frac{6}{R^6}\frac{1}{4}\sum_i\sum_j\alpha_i\alpha_j\frac{\varepsilon_i\varepsilon_j}{\varepsilon_i + \varepsilon_j} = -\frac{6}{R^6}C_{11}
\end{aligned}
\tag{11.29}
$$

where

$$C_{11} = \frac{1}{4} \sum_i \sum_j \alpha_i \alpha_j \frac{\varepsilon_i \varepsilon_j}{\varepsilon_i + \varepsilon_j} \qquad (11.30)$$

is the dipole *dispersion constant*, the typical quantum mechanical part of the calculation of the dispersion coefficient, while 6 is a geometrical factor[2]. Therefore:

$$C_6 = 6C_{11} \qquad (11.31)$$

is the C_6 London dispersion coefficient for the long-range interaction between two ground-state H atoms.

In this way, the previously calculated *dipole pseudospectra* $\{\alpha_i, \varepsilon_i\}$, $i = 1, 2, \ldots, N$, for *each* H atom can be used to obtain better and better values for the C_6 London dispersion coefficient for the H–H interaction: a molecular (two-centre) quantity, C_6, can be evaluated in terms of atomic (one-centre), *nonobservable*, quantities, α_i (α alone is useless). The coupling between the different components of the polarizabilities occurs through the denominator in the London formula (11.30), so that we cannot sum over i or j to get the full, *observable*,[3] α^A or α^B. An alternative, yet equivalent, formula for the dispersion constant is due to Casimir and Polder (1948) in terms of the frequency-dependent polarizabilities at imaginary frequencies of A and B:

$$
\begin{cases}
C_{11} = \dfrac{1}{2\pi} \displaystyle\int_0^\infty \mathrm{d}u \, \alpha^A(iu) \, \alpha^B(iu) \\[2mm]
\alpha^A(iu) = \displaystyle\sum_\kappa \varepsilon_\kappa \frac{2\mu(0\kappa)\mu(\kappa 0)}{\varepsilon_\kappa^2 + u^2}, \quad \alpha^A \text{ (static)} = \alpha^A(0) = \lim_{u \to 0} \alpha^A(iu)
\end{cases}
$$

$$(11.32)$$

where u is a *real* quantity.

---

[2] Depending on the spherical symmetry of the ground-state H atoms.

[3] That is, measurable.

Table 11.1 N-term results for the dipole dispersion constant C_{11} and C_6 London dispersion coefficients for the H–H interaction

N	$C_{11}/E_h a_0^6$	$C_6/E_h a_0^6$	Accurate/%
1	1	6	92.3
2	1.080 357	6.482 1	99.7
3	1.083 067	6.498 4	99.99
4	1.083 167	6.499 00	99.999
5	1.083 170	6.499 02	100

In this case, we must know the dependence of the frequency-dependent polarizabilities on the real frequency u, and the coupling occurs now via the integration over the frequencies. When the necessary data are available, however, the London formula (11.30) is preferable because use of the Casimir–Polder formula (11.32) presents some problems in the accurate evaluation of the integral through numerical quadrature techniques (Figari and Magnasco, 2003).

Using the London formula and some of the pseudospectra derived in Chapter 10, we obtain for the leading term of the H–H interaction the results collected in Table 11.1.

Table 11.1 shows that convergence is very rapid for the H–H interaction. We give here the explicit calculation for $N = 2$:

$$
C_{11}(\text{2-term}) = \frac{1}{8}\left(\frac{25}{6}\right)^2 \frac{2}{5} + \frac{1}{8}\left(\frac{2}{6}\right)^2 \times 1 + \frac{\frac{1}{2}\frac{25}{6}\frac{2}{6}\frac{2}{5} \times 1}{\frac{2}{6}\frac{2}{6}\frac{2}{5} + 1}
$$

$$
= \frac{5 \times 5 \times 25 \times 2}{2 \times 4 \times 36 \times 5} + \frac{4}{2 \times 4 \times 36} + \frac{5 \times 5 \times 2 \times 5}{36 \times \cdot 5 \times 7}
$$

$$
= \frac{125}{4 \times 36} + \frac{1}{2 \times 36} + \frac{50}{7 \times 36} = \frac{1089}{1008} = \frac{121}{112} = 1.080\,357
$$

so that the two-term approximation gives the dispersion constant as the ratio between two not divisible integers! However, this explicit calculation is no longer possible for $N > 2$, where we must resort to the numerical methods touched upon in Chapter 10. Using a nonvariational technique in momentum space, Koga and Matsumoto (1985) gave the three-term C_6 for H–H as the ratio of not divisible integers as

$$
C_6 = \frac{12\,529}{1928} = 6.498\,443\,983\ldots
$$

and for the four-term C_6

$$C_6 = \frac{6\,313\,807}{971\,504} = 6.499\,002\,577\ldots$$

This last value is accurate to four decimal figures,[4] while a value accurate to 13 decimal figures was given by Thakkar (1988):

$$C_6 = 6.499\,026\,705\,405\,840\,5\ldots$$

The London C_6 dispersion coefficient for the long-range H–H interaction is today one of the best known 'benchmarks' in the Literature. It was calculated with an accuracy of 20 exact decimal digits by Yan *et al.* (1996):

$$C_6 = 6.499\,026\,705\,405\,839\,313\,13\ldots$$

and, with an accuracy of 15 decimal digits, in an independent way, by Koga and Matsumoto (1985) and by Magnasco *et al.* (1998):

$$C_6 = 6.499\,026\,705\,405\,839\,\mathbf{218}\ldots$$

It must be stressed that all such values are far beyond any possible experimental accuracy, being useful only for checking the accuracy of different ways of calculation!

Unfortunately, the convergence rate for C_6 (as well as that for α) is not so good for two-electron systems, as we shall see shortly.

11.3   MOLECULAR INTERACTIONS

The nonexpanded intermolecular potential V arises from the Coulombic interactions between all pairs i, j of charged particles (electrons plus nuclei) in the molecules (Figure 11.3):

$$V = \sum_i \sum_j \frac{q_i q_j}{r_{ij}} \tag{11.33}$$

---

[4] Inaccurate digits are in bold type.

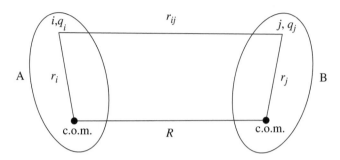

Figure 11.3 Interparticle distances in the intermolecular potential (c.o.m.: centre of mass)

where q_i and q_j are the charges of particles i (belonging to A) and j (belonging to B) interacting at the distance r_{ij}.

11.3.1 Nonexpanded Energy Corrections up to Second Order

If A_0 and B_0 are the unperturbed wavefunctions of molecules A (N_A electrons) and B (N_B electrons), and A_i and B_j a pair of excited pseudostates describing single excitations on A and B, all fully antisymmetrized within the space of A and B, we have to second order of RS perturbation theory that

$$E_1^{cb} = \langle A_0 B_0 | V | A_0 B_0 \rangle = E_1^{es} \tag{11.34}$$

is the semiclassical electrostatic energy arising in first order from the interactions between *undistorted* A and B; that

$$\tilde{E}_2^{ind,A} = -\sum_i \frac{|\langle A_i B_0 | V | A_0 B_0 \rangle|^2}{\varepsilon_i} = -\sum_i \frac{|(A_0 A_i | U^B)|^2}{\varepsilon_i} \tag{11.35}$$

is the polarization (distortion) of A by the static field of B, described by

$$U^B = \langle B_0 | V | B_0 \rangle \tag{11.36}$$

the MEP of B; that

$$\tilde{E}_2^{\text{ind},B} = -\sum_j \frac{|\langle A_0 B_j | V | A_0 B_0 \rangle|^2}{\varepsilon_j} = -\sum_j \frac{|(B_0 B_j | U^A)|^2}{\varepsilon_j} \qquad (11.37)$$

is the polarization (distortion) of B by the static field of A, described by the MEP U^A; and that

$$\tilde{E}_2^{\text{disp}} = -\sum_i \sum_j \frac{|\langle A_i B_j | V | A_0 B_0 \rangle|^2}{\varepsilon_i + \varepsilon_j} = -\sum_i \sum_j \frac{\left|\left\langle A_i B_j \left| \sum_{i' < j'} r_{i'j'}^{-1} \right| A_0 B_0 \right\rangle\right|^2}{\varepsilon_i + \varepsilon_j}$$

$$(11.38)$$

is the dispersion interaction, a purely electronic term arising from the *density fluctuations* of the electrons on A and B which are coupled together through the intermolecular electron repulsion operator r_{12}^{-1} (1 on A, 2 on B).

Generalization of the previous H–H results to *molecules* is possible in terms of the charge-density operator (Longuet-Higgins, 1956) and of static and transition electron densities, $P^A(00|\mathbf{r}_1;\mathbf{r}_1)$ and $P^B(00|\mathbf{r}_2;\mathbf{r}_2)$, $P^A(0i|\mathbf{r}_1;\mathbf{r}_1)$ and $P^B(0j|\mathbf{r}_2;\mathbf{r}_2)$ respectively on A and B. The nonexpanded dispersion energy between molecules A and B then takes the simple integral form

$$\tilde{E}_2^{\text{disp}} = -\sum_i \sum_j \frac{\left| \iint d\mathbf{r}_1\, d\mathbf{r}_2\, \dfrac{P^A(0i|\mathbf{r}_1;\mathbf{r}_1) P^B(0j|\mathbf{r}_2;\mathbf{r}_2)}{r_{12}} \right|^2}{\varepsilon_i + \varepsilon_j} \qquad (11.39)$$

which can be compared with the corresponding integral expression (11.16) found for the H–H interaction, where $P^A(0i|\mathbf{r}_1;\mathbf{r}_1) = a_0(\mathbf{r}_1)a_i^*(\mathbf{r}_1)$ and $P^B(0j|\mathbf{r}_2;\mathbf{r}_2) = b_0(\mathbf{r}_2)b_j^*(\mathbf{r}_2)$.

11.3.2 Expanded Energy Corrections up to Second Order

In molecules, the interaction depends on the distance R between their centres of mass as well as on the relative orientation of the interacting partners, which can be specified in terms of the five independent angles[5]

---

[5] These angles are simply related to the Euler angles describing the rotation of a rigid body (Brink and Satchler, 1993).

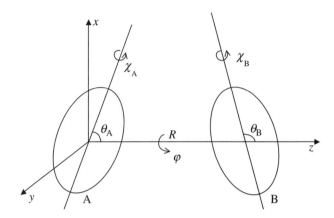

Figure 11.4 The five angles specifying in general the relative orientation of two polyatomic molecules

$(\theta_A, \theta_B, \varphi, \chi_A, \chi_B)$ shown in Figure 11.4. The first three angles describe the orientation of the principal symmetry axes of the two molecules and the latter two the rotation about these axes.

In what follows, we shall limit ourselves mostly to consideration of the long-range dispersion interaction between (i) two linear molecules A and B (top of Figure 11.5) and (ii) an atom A, at the origin of the intermolecular coordinate system, and a linear molecule B, whose orientation with respect to the z-axis is specified by the single angle θ (bottom of Figure 11.5).

The linear molecule has *two* dipole polarizabilities, $\alpha^{\|}$, the parallel or longitudinal component directed along the intermolecular axis, and $\alpha^{\perp}$, the perpendicular or transverse component perpendicular to the inter-molecular axis (McLean and Yoshimine, 1967a). The molecular *isotropic* polarizability can be compared to that of atoms, and is defined as

$$\alpha = \frac{\alpha^{\|} + 2\alpha^{\perp}}{3} \tag{11.40}$$

while

$$\Delta\alpha = \alpha^{\|} - \alpha^{\perp} \tag{11.41}$$

is the polarizability *anisotropy*, which is zero for $\alpha^{\perp} = \alpha^{\|}$.

The composite system of two *different* linear molecules hence has four independent elementary dipole dispersion constants, which in London

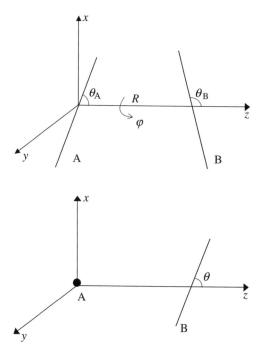

Figure 11.5 Top, the three angles specifying the relative orientation of two linear molecules. Bottom: the system atom A–linear molecule B

form can be written as

$$
\begin{cases}
A = \dfrac{1}{4}\sum_i \sum_j \alpha_i^{\|} \alpha_j^{\|}\, \dfrac{\varepsilon_i^{\|}\varepsilon_j^{\|}}{\varepsilon_i^{\|}+\varepsilon_j^{\|}}, &
B = \dfrac{1}{4}\sum_i \sum_j \alpha_i^{\|} \alpha_j^{\perp}\, \dfrac{\varepsilon_i^{\|}\varepsilon_j^{\perp}}{\varepsilon_i^{\|}+\varepsilon_j^{\perp}}, \\[4mm]
C = \dfrac{1}{4}\sum_i \sum_j \alpha_i^{\perp} \alpha_j^{\|}\, \dfrac{\varepsilon_i^{\perp}\varepsilon_j^{\|}}{\varepsilon_i^{\perp}+\varepsilon_j^{\|}}, &
D = \dfrac{1}{4}\sum_i \sum_j \alpha_i^{\perp} \alpha_j^{\perp}\, \dfrac{\varepsilon_i^{\perp}\varepsilon_j^{\perp}}{\varepsilon_i^{\perp}+\varepsilon_j^{\perp}}
\end{cases}
\qquad (11.42)
$$

For two *identical* linear molecules, there are three independent dispersion constants, since $C = B$.

Spherical tensor expansion of the product $r_{12}^{-1} \cdot r_{1'2'}^{-1}$ (Wormer, 1975; Magnasco and Ottonelli, 1999a) allows us to write the leading (dipole–dipole) term of the long-range dispersion interaction between two linear molecules in the form:

$$
\tilde{E}_2^{\text{disp}} = -R^{-6}C_6(\theta_A, \theta_B, \varphi) \qquad (11.43)
$$

$C_6(\theta_A, \theta_B, \varphi)$ being an *angle-dependent* dipole dispersion coefficient, which can be expressed (Meyer, 1976) in terms of associated Legendre polynomials on A and B as

$$C_6(\theta_A, \theta_B, \varphi) = C_6 \sum_{L_A L_B M} \gamma_6^{L_A L_B M} P_{L_A}^M (\cos \theta_A) P_{L_B}^M (\cos \theta_B) \qquad (11.44)$$

where L_A, $L_B = 0, 2$ and $M = |M| = 0, 1, 2$. In (11.44), C_6 is the isotropic coefficient and γ_6 an anisotropy coefficient, defined as

$$\gamma_6^{L_A L_B M} = \frac{C_6^{L_A L_B M}}{C_6} \qquad (11.45)$$

The different components of the C_6 dispersion coefficients in the $L_A L_B M$ scheme for (i) two *different* linear molecules and (ii) an atom and a linear molecule are given in Table 11.2 (Magnasco and Ottonelli, 1999a) in terms of the symmetry-adapted combinations of the elementary dispersion constants (11.42). The coefficients with $M \neq 0$ are not independent, but are related to that with $M = 0$ by the relations

$$C_6^{221} = -\frac{2}{9} C_6^{220}, \quad C_6^{222} = \frac{1}{36} C_6^{220} \qquad (11.46)$$

For *identical* molecules, $C = B$ in (11.42) and the (020) and (200) coefficients are equal.

Table 11.2 $L_A L_B M$ components of C_6 dispersion coefficients for (i) two linear molecules and (ii) an atom and a linear molecule

$L_A L_B M$	(i)	(ii)
0 0 0	$\frac{2}{3}(A + 2B + 2C + 4D)$	$2A + 4B$
0 2 0	$\frac{2}{3}(A - B + 2C - 2D)$	$2A - 2B$
2 0 0	$\frac{2}{3}(A + 2B - C - 2D)$	
2 2 0	$2(A - B - C + D)$	
2 2 1	$-\frac{4}{9}(A - B - C + D)$	
2 2 2	$\frac{1}{18}(A - B - C + D)$	

Therefore, the determination of the elementary dispersion constants (the quantum mechanical relevant part of the calculation) allows for a detailed analysis of the angle-dependent dispersion coefficients between molecules.

An equivalent, yet explicit, expression of the C_6 angle-dependent dispersion coefficient for the homodimer of two linear molecules as a function of the three independent dispersion constants was derived by Briggs *et al.* (1971) in their attempt to determine the dispersion coefficients of two H_2 molecules in terms of nonlinear $^1\Sigma_u^+$ and $^1\Pi_u$ pseudostates:

$$C_6\,(\theta_A, \theta_B, \varphi) = (2B + 4D) + 3(B - D)(\cos^2\theta_A + \cos^2\theta_B)$$

$$+ (A - 2B + D)(\sin\theta_A \sin\theta_B \cos\varphi - 2\cos\theta_A \cos\theta_B)^2 \tag{11.47}$$

Since $(\cos\theta = x)$

$$\langle\cos^2\theta\rangle = \frac{\displaystyle\int_{-1}^{1} dx\, x^2}{\displaystyle\int_{-1}^{1} dx} = \frac{1}{3}, \quad \langle\sin^2\theta\rangle = \frac{2}{3}, \quad \langle\cos^2\varphi\rangle = \frac{\displaystyle\int_{0}^{2\pi} d\varphi \cos^2\varphi}{\displaystyle\int_{0}^{2\pi} d\varphi} = \frac{1}{2} \tag{11.48}$$

averaging (11.47) over angles and noting that only *squared* terms contribute to the average, we obtain the following for the *isotropic* C_6 dispersion coefficient:

$$\langle C_6\rangle = (2B + 4D) + 3(B - D)\left(\frac{1}{3} + \frac{1}{3}\right) + (A - 2B + D)\left(\frac{2}{3}\times\frac{2}{3}\times\frac{1}{2} + 4\times\frac{1}{3}\times\frac{1}{3}\right)$$

$$= \frac{2}{3}(A + 4B + 4D) = C_6 \tag{11.49}$$

in accord with the result of the first row of Table 11.2. Magnasco *et al.* (1990b) gave an alternative interesting expression for $C_6(\theta_A, \theta_B, \varphi)$ in terms of frequency-dependent *isotropic polarizabilities* $\alpha(iu)$ and *polarizability anisotropies* $\Delta\alpha(iu)$ of the two linear molecules:

$$C_6(\theta_A, \theta_B, \varphi) = \frac{1}{2\pi}\int_0^\infty du \left\{ \alpha^A(iu)\alpha^B(iu) + [(3\cos^2\theta_B - 1)\alpha_A(iu)\Delta\alpha_B(iu) \right.$$

$$+ (3\cos^2\theta_A - 1)\Delta\alpha_A(iu)\alpha_B(iu)]$$

$$+ [4\cos^2\theta_A\cos^2\theta_B - \cos^2\theta_A - \cos^2\theta_B - \sin^2\theta_A\sin^2\theta_B\cos\varphi$$

$$+ \sin^2\theta_A\sin^2\theta_B\cos^2\varphi]\Delta\alpha_A(iu)\Delta\alpha_B(iu) \right\} \tag{11.50}$$

Averaging over angles, all coefficients involving polarizability anisotropies are zero, giving the isotropic C_6 coefficient in the Casimir–Polder form (11.32).

To get an illustrative numerical example, we can use the four-term pseudospectrum of Table 11.3 for the dipole polarizabilities of the $^1\Sigma_g^+$ ground state of the H_2 molecule at $R = 1.4\, a_0$, which gives for the dipole polarizabilities of H_2

$$\alpha^\| = 6.378, \quad \alpha^\perp = 4.559 \tag{11.51}$$

results that are remarkably good for both polarizabilities, being 99.9% and 99.6% respectively of the accurate values ($\alpha^\| = 6.383$ and $\alpha^\perp = 4.577$) obtained from the accurate 34-term pseudospectrum (Magnasco and Ottonelli, 1996a).

We can then calculate the four-term approximation to the three independent elementary dispersion constants (11.42) for the *homodimer* H_2–H_2, obtaining the following numerical results:

$$A = 2.683 \qquad 99.8\% \text{ of the accurate value } 2.689$$
$$B = C = 2.018 \qquad 99.3\% \text{ of the accurate value } 2.032$$
$$D = 1.524 \qquad 98.8\% \text{ of the accurate value } 1.542$$

Table 11.3   Four-term dipole pseudospectrum (atomic units) of H_2 (Σ_g^+) at $R = 1.4a_0$

i	$\alpha_i^\|$	$\varepsilon_i^\|$	$\alpha_i^\perp$	$\varepsilon_i^\perp$
1	4.567	0.473	2.852	0.494
2	1.481	0.645	1.350	0.699
3	0.319	0.973	0.335	1.157
4	0.011	1.701	0.022	2.207

From these values, we obtain for the isotropic C_6 dispersion coefficient for H_2–H_2

$$C_6 = C_6^{000} = 11.23 E_h a_0^6$$

and for the first dipole anisotropy we have

$$\gamma_0^{020} = \frac{C_6^{020}}{C_6} = 0.098$$

which are respectively within 99.2% and +2% of the accurate values ($C_6 = 11.32$ and $\gamma_6 = 0.096$) given by Magnasco and Ottonelli (1996a).

As a second example, illustrating a *heterodimer* calculation, consider the C_6 dispersion coefficient of the H–H_2 system (atom–linear molecule interaction):

$$C_6(\theta) = C_6[1 + \gamma_6 P_2(\cos\theta)] \qquad (11.52)$$

where

$$P_2(\cos\theta) = \frac{3\cos^2\theta - 1}{2} \qquad (11.53)$$

is the Legendre polynomial of degree 2 (Chapter 3), C_6 is the isotropic coefficient and γ_6 is the anisotropy coefficient. From Table 11.2:

$$C_6 = C_6^{000} = 2A + 4B, \quad \gamma_6^{020} = \frac{C_6^{020}}{C_6} = \frac{A - B}{A + 2B}$$

so that, using the four-term pseudospectrum for the H atom (Table 10.2) and the one for the H_2 molecule (Table 11.3), we obtain $A = 1.696$ and $B = 1.269$ and for the isotropic C_6 dispersion coefficient of the H–H_2 interaction we obtain

$$C_6 = 2 \times 1.696 + 4 \times 1.269 = 8.468 E_h a_0^6$$

which is within 99.6% of the accurate value $C_6 = 8.502$ (Magnasco, Ottonelli, 1996b). The calculated value of $\gamma_6 = 0.101$ exceeds the correct

Table 11.4 N-term convergence of ground-state isotropic dipole polarizabilities α (a_0^3) for simple atoms and molecules

N	H	He	$H_2^{+\,a}$	$H_2^{\,b}$
1	4	0.694	1.485	2.523
2	4.5	1.042	2.614	3.481
3		1.078	2.819	3.493
4		1.082	2.836	4.078
5		1.135	2.837	4.452
10		1.364	2.855	4.999
15		1.378	2.864	5.145
20		1.382	2.864	5.165
Accurate	4.5^c	1.383^d	2.864^e	5.181^f

[a] H_2^+: $R_e = 2a_0$.
[b] H_2: $R_e = 1.4a_0$.
[c] Exact value.
[d] Yan *et al.*, 1996.
[e] Bishop and Cheung, 1978; Babb, 1994; Magnasco and Ottonelli, 1999b.
[f] Bishop *et al.*, 1991.

value (0.099) by 2%. So, these homo- and hetero-dimer calculations are affected essentially by the same percent errors[6].

Tables 11.4 and 11.5 show, respectively, the convergence of N-term dipole pseudospectra in the calculation of polarizabilities α and isotropic

Table 11.5 N-term convergence of isotropic C_6 dispersion coefficients $(E_h a_0^6)$ for the homodimers of simple atoms and molecules

N	H–H	He–He	$H_2^+ - H_2^{+\,a}$	$H_2 - H_2^{\,b}$
1	6	0.695	1.449	4.729
2	6.482	1.142	2.965	7.325
3	6.498	1.186	3.218	7.364
4	6.499	1.189	3.234	8.825
5	6.499	1.242	3.234	9.645
10		1.443	3.265	10.965
15		1.455	3.284	11.263
20		1.459	3.284	11.289
Accurate	6.499^c	1.461^d	3.284^e	11.324^f

[a] H_2^+: $R_e = 2a_0$.
[b] H_2: $R_e = 1.4a_0$.
[c] Exact value.
[d] Yan *et al.*, 1996.
[e] Babb, 1994; Magnasco and Ottonelli, 1999b.
[f] Magnasco and Ottonelli, 1996a.

---

[6] These errors are completely removed when using *best* four-term reduced pseudospectra for H and H_2 obtained from a recently derived efficient interpolation technique (Figari *et al.*, 2007).

C_6 dispersion coefficients for the homodimers of some simple one-electron and two-electron atomic and molecular systems. It is seen that the convergence rate of either α or C_6 is remarkably slower for the two-electron systems.

We see from Table 11.4 that the five-term approximation gives 82% for He, 99% for H_2^+ and 86% for H_2, showing that convergence is sensibly worst for the two-electron systems. Table 11.5 shows that the five-term approximation, yielding practically the exact result for H–H, gives 85% for He–He and H_2–H_2 and over 98% for H_2^+ –H_2^+ , confirming the similar results for polarizabilities.

11.3.3   Other Expanded Interactions

At variance with the dispersion interaction, whose calculation at long range requires knowledge of N-term pseudospectra of the monomer molecules, which are *not observable* quantities, the remaining components of the intermolecular interaction, the electrostatic and induction energies, are instead expressible in terms of physically *observable* electric properties of the interacting molecules, namely the permanent moments μ and the total polarizabilities α of the individual molecules.

(a) As an example, the leading term of the electrostatic interaction between two dipolar $HF(^1\Sigma^+)$ molecules[7] is given at long range by the dipole–dipole interaction[8] going as R^{-3}:

$$E_1^{es} = \frac{\mu_{HF}^2}{R^3}\left(\sin\theta_A\sin\theta_B\cos\varphi - 2\cos\theta_A\cos\theta_B\right) \qquad (11.54)$$

which has a minimum for the head-to-tail configuration of the two molecules:

$$E_1^{es} = -2\frac{\mu_{HF}^2}{R^3} \qquad (11.55)$$

for $\theta_A = \theta_B = 180°$, leading to the formation of a collinear H-bond (H–F $\cdots$ H–F). The next quadrupole–dipole and dipole–quadrupole

---

[7] Namely, molecules whose first nonzero permanent moment is the dipole moment.

[8] Equation 11.54 is just Equation 11.22 weighted in first order with the ground-state wavefunction when the resulting dipole components are expressed in spherical coordinates.

Figure 11.6 Experimentally observed structure for $(HF)_2$

interaction terms, going as R^{-4}, favour instead an L configuration of the dimer, giving as a whole the structure of the homodimer with $\theta \approx 60°$ observed by experiment and schematically represented in Figure 11.6.

A similar result is obtained by the qualitative MO description of the H-bond in $(HF)_2$ depicted in the drawings of Figure 11.7, which show electron transfer from F lone pairs (doubly occupied MOs) to a vacant (empty MO) orbital on H. The top row shows σ charge transfer from F to H, the bottom row π charge transfer from F to H, which must be doubled on account of the twofold degeneracy of the π energy level (the figure shows the π_x MO). As a result, the dimer roughly assumes the noncollinear geometrical structure[9] depicted in Figure 11.6.

(b) As a further example, the leading R^{-6} term describing the polarization of the He atom by the HF molecule, namely

$$E_2^{\text{ind}}(6) = -\frac{\alpha_{\text{He}}\mu_{\text{HF}}^2}{R^6}\frac{3\cos^2\theta + 1}{2} \qquad (11.56)$$

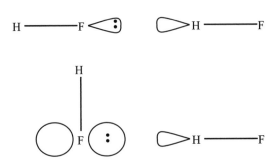

Figure 11.7 MO description of the origin of the H-bond in $(HF)_2$

---

[9] It must be remarked that molecular beam experimentalists refer to the geometrical shape of Figure 11.6 as a *linear* dimer.

is unable to discriminate between H-bonded ($\theta = 180°$) and anti-H-bonded ($\theta = 0°$) configurations, giving in both cases

$$E_2^{ind}(6) = -2\frac{\alpha_{He}\mu_{HF}^2}{R^6} \qquad (11.57)$$

The next term in R^{-7}, which implies further polarization of the He atom by the *mixed* dipole–quadrupole moments of HF, contains a $\cos^3\theta$ term:[10]

$$E_2^{ind}(7) = \frac{3\alpha_{He}\mu_{HF}\Theta_{HF}}{R^7}\cos^3\theta \qquad (11.58)$$

which stabilizes the H-bonded configuration He $\cdots$ H–F ($\theta = 180°$), so that we can appropriately speak of formation of an H-bond between He and HF (Magnasco *et al.*, 1989).

11.4   VAN DER WAALS AND HYDROGEN BONDS

From all we have seen so far, we can say that a *VdW bond* occurs when the small Pauli repulsion arising from the first-order interaction of closed-shell molecules at long range is overbalanced by weak attractive second-order induction and dispersion forces. VdW molecules are weakly bound complexes with large-amplitude vibrational structure (Buckingham, 1982). This is the case of the dimers of the rare gases X_2 ($X =$ He, Ne, Ar, Kr, Xe) or the weak complexes between centrosymmetrical molecules like $(H_2)_2$ or $(N_2)_2$. Complexes between proton-donor and proton-acceptor molecules, like $(HF)_2$ or $(H_2O)_2$, involve formation of hydrogen bonds (H-bonds), which are essentially electrostatic in nature, and lie on the borderline between VdW molecules and 'good' molecules, having sensibly larger intermolecular energies.

Figure 11.8 shows the VdW potential curve resulting for the interaction of two ground-state He atoms at medium range. The upper curve is the first-order energy E_1 mostly arising from repulsive (Pauli) exchange-overlap; the bottom curve is the resultant of adding the attractive energy $\tilde{E}_2^{disp}$ due to London dispersion. A weak potential minimum (about $33 \times 10^{-6}E_h$) is observed at the rather large distance of $R_e = 5.6a_0$. It

---

[10] Θ is the permanent quadrupole moment of HF.

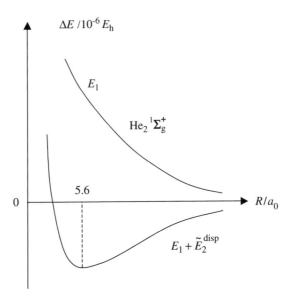

Figure 11.8   Origin of the VdW bond in He_2 $(^1\Sigma_g^+)$

should be remarked that, in this region of internuclear distances, both electrostatic and induction energies are negligible because of their charge-overlap nature. For simple 1s STOs on each He atom, with orbital exponent c_0, in fact $(\rho = c_0 R)$

$$\Delta E^{cb} = E_1^{es} = 4c_0 \frac{\exp(-2\rho)}{\rho}\left(1 + \frac{5}{8}\rho - \frac{3}{4}\rho^2 - \frac{1}{6}\rho^3\right) \qquad (11.59)$$

$$\tilde{E}_2^{ind} = \tilde{E}_2^{ind,A} + \tilde{E}_2^{ind,B} = -2\sum_i \frac{|(A_0A_i|U^B)|^2}{\varepsilon_i} \qquad (11.60)$$

and, for the neutral atom, U^B decreases exponentially far from the B nucleus.

The situation is different for the long-range interaction of two ground-state H_2O molecules, whose potential energy curve is schematically shown in Figure 11.9. In this case, the first-order interaction E_1 shows in the medium range a minimum mostly arising from the dipole–dipole interaction going as R^{-3}, and the second-order interaction simply deepens this minimum, strengthening the bond. As already seen in the case of $(HF)_2$, it is appropriate in this case to speak of formation of a *hydrogen bond*, essentially electrostatic in origin. It is of interest to note the change

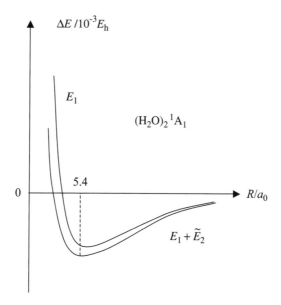

Figure 11.9   Origin of the hydrogen bond in $(H_2O)_2$ $(^1A_1)$

in the scale factor for energy, from $10^{-6}E_h$ for He_2 (VdW bond) to $10^{-3}E_h$ for $(H_2O)_2$ (H-bond), even though roughly in the same region of intermolecular distances ($5.6a_0$ for He_2 and $5.4a_0$ for the H_2O dimer).

On these grounds, some time ago, Magnasco *et al.* (1990a) derived a simple *electrostatic model* for VdW complexes, where the angular geometry of the dimers was predicted in terms of just the first two observable electric moments of the monomers. The model allowed for the successful prediction of the most stable angular shapes of 35 VdW dimers.

11.5   THE KEESOM INTERACTION

Keesom (1921) pointed out that if two dipolar molecules undergo thermal motions, then they will on average assume orientations leading to *attraction* according to

$$E_6(\text{Keesom}) = -\frac{C_6(T)}{R^6} \qquad (11.61)$$

$C_6(T)$ being the T-dependent coefficient:

$$C_6(T) = \frac{2\mu_A^2\mu_B^2}{3kT} \qquad (11.62)$$

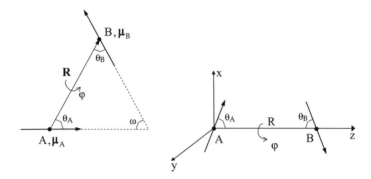

Figure 11.10   Different coordinate systems for two interacting dipoles

where T is the absolute temperature and k the Boltzmann constant. This can be explained as follows.

With reference to Figure 11.10, let us first give some alternative expressions for the interaction between dipoles (Coulson, 1958):

$$
\begin{aligned}
V &= \frac{\boldsymbol{\mu}_A \cdot \boldsymbol{\mu}_B}{R^3} - 3 \frac{(\boldsymbol{\mu}_A \cdot \mathbf{R})(\boldsymbol{\mu}_B \cdot \mathbf{R})}{R^5} \\[4pt]
&= \frac{\mu_A \mu_B}{R^3} \cos \omega - 3 \frac{(\mu_A R \cos \theta_A)(\mu_B R \cos \theta_B)}{R^5} \\[4pt]
&= \frac{\mu_A \mu_B}{R^3} (\cos \omega - 3 \cos \theta_A \cos \theta_B) \\[4pt]
&= \frac{\mu_A \mu_B}{R^3} (\sin \theta_A \sin \theta_B \cos \varphi - 2 \cos \theta_A \cos \theta_B)
\end{aligned}
\tag{11.63}
$$

since, by the addition theorem (MacRobert, 1947):

$$
\cos \omega = \cos \theta_A \cos \theta_B + \sin \theta_A \sin \theta_B \cos \varphi
\tag{11.64}
$$

where $\varphi = \varphi_A - \varphi_B$ is the dihedral angle between the planes specified by μ_A, μ_B and R. The last expression in (11.63) is the most convenient for us, giving the dipole interaction in terms of the spherical coordinates R, θ_A, θ_B and φ.

It is convenient to put

$$
\begin{cases}
\Omega = \theta_A, \theta_B, \varphi \\
F(\Omega) = \sin \theta_A \sin \theta_B \cos \varphi - 2 \cos \theta_A \cos \theta_B
\end{cases}
\tag{11.65}
$$

If all orientations were *equally probable*, then the average potential energy $\langle V \rangle$, and hence the average first-order electrostatic $\langle C_3 \rangle$ coefficient (Magnasco *et al.*, 1988; Magnasco, 2007), would be zero. In fact:

$$\langle V \rangle_\Omega = \frac{\mu_A \mu_B}{R^3} \frac{\int_\Omega d\Omega \, F(\Omega)}{\int_\Omega d\Omega} = 0 \qquad (11.66)$$

since

$$\int_\Omega d\Omega = \int_0^{2\pi} d\varphi \int_0^\pi d\theta_A \sin\theta_A \int_0^\pi d\theta_B \sin\theta_B = 2\pi \int_{-1}^1 dx_A \int_{-1}^1 dx_B = 8\pi \qquad (11.67)$$

where

$$x_A = \cos\theta_A, \quad x_B = \cos\theta_B \qquad (11.68)$$

$$\int_\Omega d\Omega \, F(\Omega) = \int_0^{2\pi} d\varphi \cos\varphi \left[\int_{-1}^1 dx \,(1-x^2)^{1/2} \right]^2 - 2 \int_0^{2\pi} d\varphi \left(\int_{-1}^1 dx \, x \right)^2 = 0 \qquad (11.69)$$

The vanishing of the average potential energy for *free* orientations is true for all multipoles (dipoles, quadrupoles, octupoles, hexadecapoles, etc.).

The Boltzmann probability for a dipole arrangement whose potential energy is V is instead proportional to

$$W \propto \exp(-V/kT) \qquad (11.70)$$

We now average the quantity $V\exp(-V/kT)$ over all possible orientations Ω assumed by the dipoles:

$$\langle V\exp(-V/kT) \rangle = \frac{\mu_A \mu_B}{R^3} \frac{\int_\Omega d\Omega \, F(\Omega)\exp[aF(\Omega)]}{\int_\Omega d\Omega \, \exp[aF(\Omega)]} \qquad (11.71)$$

where we have introduced the T-dependent negative quantity

$$a = - \frac{\mu_A \mu_B}{R^3 kT} < 0 \qquad (11.72)$$

We then obtain the familiar formula[11]

$$\langle V \exp(-V/kT) \rangle = \frac{\mu_A \mu_B}{R^3} \frac{\int_\Omega d\Omega\, F(\Omega) \exp[aF(\Omega)]}{\int_\Omega d\Omega\, \exp[aF(\Omega)]} = \frac{\mu_A \mu_B}{R^3} \frac{d}{da} \ln K(a) \qquad (11.73)$$

where

$$K(a) = \int_\Omega d\Omega\, \exp[aF(\Omega)] \qquad (11.74)$$

is called the Keesom integral.

We evaluate (11.73) for $a \approx$ *small* (high temperatures and large distances between the dipoles) by expanding the exponential in (11.74):

$$\int_\Omega d\Omega\, \exp[aF(\Omega)] \approx \int_\Omega d\Omega \left[1 + aF(\Omega) + \frac{a^2}{2} F(\Omega)^2 + \cdots \right] \qquad (11.75)$$

where we have just seen that, in the expansion, the second integral[12] vanishes, so that only the quadratic term contributes to the Keesom integral.

We have

$$\int_\Omega d\Omega\, F(\Omega)^2 = \int_0^{2\pi} d\varphi \int_0^\pi d\theta_A \sin\theta_A \int_0^\pi d\theta_B \sin\theta_B$$

$$\times (\sin^2\theta_A \sin^2\theta_B \cos^2\varphi + 4\cos^2\theta_A \cos^2\theta_B)$$

---

[11] Relation (11.73) is much the same as that observed in the Debye theory of the orientation of electric dipoles in gases, and in the Langevin (classical) or Brillouin (quantal) equations for the paramagnetic gas. In the last case, summations replace integrations over the parameter a.

[12] And all *odd* powers of it.

$$= \int_0^{2\pi} d\varphi \cos^2 \varphi \int_{-1}^1 dx_A \, (1 - x_A^2) \int_{-1}^1 dx_B \, (1 - x_B^2)$$

$$+ 4 \int_0^{2\pi} d\varphi \int_{-1}^1 dx_A \, x_A^2 \int_{-1}^1 dx_B \, x_B^2 = 8\pi \frac{2}{3} \tag{11.76}$$

so that

$$\int_\Omega d\Omega \, \frac{a^2}{2} F(\Omega)^2 = 8\pi \frac{a^2}{3} \tag{11.77}$$

Then:

$$\int_\Omega d\Omega \left[1 + \frac{a^2}{2} F(\Omega)^2 \right] = 8\pi \left(1 + \frac{a^2}{3} \right) \tag{11.78}$$

$$\frac{d}{da} \ln \left[8\pi \left(1 + \frac{a^2}{3} \right) \right] = \frac{d}{da} \left[\ln 8\pi + \ln \left(1 + \frac{a^2}{3} \right) \right]$$

$$= \frac{1}{1 + \frac{a^2}{3}} \frac{2}{3} a \approx \frac{2}{3} a \tag{11.79}$$

for $a \approx$ *small*. Hence, we obtain the final result for the average attraction energy between the dipoles:

$$\langle V \exp(-V/kT) \rangle \approx \frac{\mu_A \mu_B}{R^3} \frac{2}{3} a = - \frac{2}{3kT} \frac{\mu_A^2 \mu_B^2}{R^6} \tag{11.80}$$

This is known as the Keesom or dipole orientation energy (Equations 11.61 and 11.62). This term depends on R^{-6}, but is temperature dependent and decreases in importance with increasing T.

It is of interest to compare the relative importance of all *attractive* contributions to the intermolecular energy in the VdW region. For atoms and centrosymmetrical molecules, induction is zero, so that the only

Table 11.6 Comparison between isotropic C_6 coefficients ($E_h a_0^6$) for some homo-dimers of atoms and molecules in the gas phase at $T = 293\,K$

Atom–atom	Dispersion	Molecule–molecule	Dispersion	Induction	Keesom
He_2	1.46	$(H_2)_2$	12.1	0	0
Ne_2	6.35	$(N_2)_2$	73.4	0	0
H_2	6.50	$(CO)_2$	81.4	0.05	0.002
Ar_2	64.9	$(NO)_2$	69.8	0.08	0.009
Kr_2	129	$(N_2O)_2$	184.9	0.19	0.017
Be_2	213	$(NH_3)_2$	89.1	9.82	81.3
Xe_2	268	$(H_2O)_2$	45.4	10.4	204
Mg_2	686	$(HF)_2$	19.0	6.3	227
Li_2	1450	$(LiH)_2$	125	299	8436

contribution comes from attractive dispersion. For dipolar molecules, induction is usually negligible with respect to dispersion except for $(LiH)_2$. The electrostatic energy is not zero when its thermal average is taken. The corresponding Keesom attractive energies (11.80) are, hence, the *isotropic* electrostatic contributions to the interaction energy and are temperature dependent. A comparison between isotropic C_6 coefficients for some homodimers at $T = 293\,K$ is given in Table 11.6. It is seen that Keesom $C_6(T)$ is negligible compared with dispersion and induction coefficients for the homodimers of CO, NO, N_2O, while for $(NH_3)_2$, $(HF)_2$ and $(H_2O)_2$ the Keesom dipole orientation forces become increasingly dominant at room temperature, so they cannot be neglected in assessing collective gas properties such as the equation of state for real gases and virial coefficients.

Magnasco *et al.* (2006) recently extended Keesom's calculations up to the R^{-10} term, showing that deviations of the Keesom approximation from the full series expansion are less important than consideration of the higher order terms in the R^{-2n} expansion of the intermolecular potential.

Battezzati and Magnasco (2004) also gave an asymptotic evaluation of the Keesom integral (11.74) for $|a| \approx large$,[13] obtaining the formula

$$K(a) \cong \frac{4\pi \exp(-2a)}{3} \frac{}{a^2}\left(1 - \frac{2}{3a}\right) \qquad (11.81)$$

---

[13] This is the case of low temperatures and small distances between the dipoles, as well as the case of $(LiH)_2$, where the interacting dipoles are very large.

which was confirmed by a recent independent calculation by Abbott (2007). The expansion in *inverse* powers of |a| now gives the pair potential energy at *low* temperatures as

$$\langle V \exp(-V/kT) \rangle \cong -2\frac{\mu_A \mu_B}{R^3} + 2kT + \frac{2}{3}\frac{R^3}{\mu_A \mu_B}(kT)^2 \qquad (11.82)$$

12

Symmetry

In what follows we give a short outline of the importance of molecular symmetry in quantum mechanics and how group theoretical techniques may be of help in assessing symmetry-adapted functions for the factorization of secular equations either in MO or VB calculations.

12.1 MOLECULAR SYMMETRY

Symmetry is a property that can be used for simplifying the study of a physical system, even without doing any effective calculation. Most simple molecules have a symmetry depending on the geometrical configuration of their nuclei, and which can be classified according to the molecular point group to which they belong, for example, for the first-row hydrides, linear HF ($C_{\infty v}$), bent H_2O (C_{2v}), pyramidal NH_3 (C_{3v}), tetrahedral CH_4 (T_d).

Of the greatest importance is the symmetry of any function for use in atomic or molecular calculations, since the fundamental *theorem of symmetry* states that the matrix element of any totally symmetric operator $\hat{O}$ (the identity $\hat{I}$, or any model Hamiltonian such as total $\hat{H}$, Fock $\hat{F}$, or Hückel H) between functions having *different* symmetries is identically zero:

$$\langle \chi_\mu | \hat{O} | \chi_\nu \rangle = 0, \quad \hat{O} = \hat{I}, \hat{H}, \hat{F}, H, \quad \nu \neq \mu \qquad (12.1)$$

Methods of Molecular Quantum Mechanics: An Introduction to Electronic Molecular Structure
Valerio Magnasco
© 2009 John Wiley & Sons, Ltd

Functions having *definite* symmetry properties can be obtained by letting suitable projection operators (projectors) act upon a function having no specific symmetry, as the following two examples show.

First, it is known from elementary mathematics that a function $f(x)$ can be classified as even or odd with respect to the interchange $x \Rightarrow -x$ (for instance, $\cos x$ or $\sin x$). An arbitrary function without any symmetry can always be expressed as a linear combination of an even and an odd function as

$$f(x) = \frac{1}{2}[f(x) + f(-x)] + \frac{1}{2}[f(x) - f(-x)] = \frac{1}{2}g(x) + \frac{1}{2}u(x) \quad (12.2)$$

where we use $g(x)$ (from German *gerade*) for the even function and $u(x)$ (German, *ungerade*) for the odd function. The operation (12.2) can be aptly called the resolution of the function $f(x)$ into its components having definite symmetry properties, and it is immediately evident that any integral like

$$\int_{-\infty}^{\infty} dx\, g(x)u(x) = 0 \quad (12.3)$$

is identically zero.[1]

Second, there is the split-shell description of the atomic or molecular two-electron problem. We have seen in Chapters 4 and 9 that an acceptable two-electron wavefunction for the ground states of He (^1S) or H_2 ($^1\Sigma_g^+$) is given by

$$\Psi = \frac{a(\mathbf{r}_1)b(\mathbf{r}_2) + b(\mathbf{r}_1)a(\mathbf{r}_2)}{\sqrt{2 + 2S^2}} \frac{1}{\sqrt{2}}[\alpha(s_1)\beta(s_2) - \beta(s_1)\alpha(s_2)] \quad (12.4)$$

where $a = 1s$, $b = 1s'$ for He (^1S) (the Eckart wavefunction), and $a = 1s_A$, $b = 1s_B$ for H_2 ($^1\Sigma_g^+$) (the HL wavefunction). Both wavefunctions are the product of a symmetrical space part by an antisymmetrical spin part. The simple product $a(\mathbf{r}_1)b(\mathbf{r}_2)$ has no symmetry properties and must be acted

---

[1] Changing x into $-x$ the integral is changed into itself with a minus sign.

upon by the appropriate symmetrizing operator $\hat{P}^s$:

$$\hat{P}^s = \frac{1}{2}(\hat{I} + \hat{P}), \quad \hat{P}^a = \frac{1}{2}(\hat{I} - \hat{P}) \tag{12.5}$$

where $\hat{P}$ is the operator that interchanges $\mathbf{r}_1$ and $\mathbf{r}_2$. It is easily seen that $\hat{P}^s$ and $\hat{P}^a$ have the typical properties of projection operators (idempotency, mutual exclusivity, completeness[2]):

$$\hat{P}^s\hat{P}^s = \hat{P}^s, \quad \hat{P}^a\hat{P}^a = \hat{P}^a, \quad \hat{P}^s\hat{P}^a = \hat{P}^a\hat{P}^s = 0, \quad \hat{P}^s + \hat{P}^a = \hat{I} \tag{12.6}$$

A symmetry operation $\hat{R}$ (reflection across a symmetry plane, positive or negative rotation[3] about a symmetry axis, roto-reflection, inversion about a centre of symmetry, etc.) is that operation that interchanges identical nuclei, and can be defined in either of two equivalent ways:

> *Active* representation, where we interchange physical points
> leaving unaltered the coordinate frame   (12.7)

or

> *Passive* representation, where we act upon axes
> and leave unaltered points   (12.8)

Even if conceptually the passive representation is to be preferred, the active representation is often more easily visualizable. We shall always refer to a symmetry operation as a change of coordinate axes.

Symmetry operations are described by linear operators $\hat{R}$[4] having as representatives in a given orthonormal basis $\boldsymbol{\chi}$ *orthogonal* matrices $\mathbf{D}(R) = \mathbf{R}$, constructed as follows:

$$\hat{R}\boldsymbol{\chi} = \boldsymbol{\chi}\,\mathbf{D}_\chi(R) \tag{12.9}$$

$$\boldsymbol{\chi}^\dagger\hat{R}\boldsymbol{\chi} = (\boldsymbol{\chi}^\dagger\boldsymbol{\chi})\mathbf{D}_\chi(R) = \mathbf{D}_\chi(R) \tag{12.10}$$

---

[2] In mathematics, also called the resolution of the identity.

[3] Positive rotations are always assumed anticlockwise and negative rotations clockwise.

[4] The operators $\hat{R}$ commute with the Hamiltonian operator $\hat{H}$, namely are constants of the motion.

If a basis χ is changed into a basis χ' by the transformation U, then the representative $D_{\chi'}(R)$ of $\hat{R}$ in the new basis is related to the representative $D_\chi(R)$ in the old basis by what is called a *similarity transformation*:

$$\chi' = \chi U \Rightarrow D_{\chi'}(R) = U^{-1}D_\chi(R)U \qquad (12.11)$$

Now, let $\hat{R}q$ be the coordinate transformation of the space point q under the symmetry operation $\hat{R}$:

$$q' = \hat{R}q, \qquad (12.12)$$

and

$$f'(q) = \hat{R}f(q) \qquad (12.13)$$

be the transformation of the function $f(q)$ under the operation $\hat{R}$. A transformation of the coordinate axes does not alter the value of the function at every point of space:

$$f'(q') = f(q) \qquad (12.14)$$

Namely:

$$\hat{R}f(\hat{R}q) = f(q) \qquad (12.15)$$

If this equation is applied to the point $R^{-1}q$, then we obtain the basic relation

$$\hat{R}f(q) = f(R^{-1}q) \qquad (12.16)$$

Namely, the function transformed under a symmetry operation is equal to the function obtained by subjecting its argument to the inverse transformation.

As an example, with reference to Figure 12.1, consider the rotation C_α^+ of the functions $(p_x\, p_y)$ whose angular part is defined as

$$p_x \propto \sin\theta\cos\varphi, \quad p_y \propto \sin\theta\sin\varphi \qquad (12.17)$$

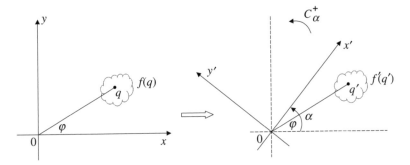

Figure 12.1 The function transformed under the positive rotation C_α^+ is equal to the function whose argument is transformed under the negative rotation C_α^-

We obtain the transformations

$$\varphi' = C_\alpha^+ \varphi = \varphi - \alpha, \quad C_\alpha^- \varphi = \varphi + \alpha \tag{12.18}$$

$$
\begin{aligned}
C_\alpha^+ p_x(\varphi) &= p_x(C_\alpha^- \varphi) = p_x(\varphi + \alpha) = \sin\theta\cos(\varphi + \alpha) \\
&= \sin\theta(\cos\varphi\cos\alpha - \sin\varphi\sin\alpha) = p_x\cos\alpha - p_y\sin\alpha
\end{aligned}
\tag{12.19}
$$

$$
\begin{aligned}
C_\alpha^+ p_y(\varphi) &= p_y(C_\alpha^- \varphi) = p_y(\varphi + \alpha) = \sin\theta\sin(\varphi + \alpha) \\
&= \sin\theta(\sin\varphi\cos\alpha + \cos\varphi\sin\alpha) = p_x\sin\alpha + p_y\cos\alpha
\end{aligned}
\tag{12.20}
$$

which can be written in matrix form as[5]

$$C_\alpha^+ (p_x\, p_y) = (C_\alpha^+ p_x C_\alpha^+ p_y) = (p_x\, p_y)\begin{pmatrix} \cos\alpha & \sin\alpha \\ -\sin\alpha & \cos\alpha \end{pmatrix} = (p_x\, p_y)\mathbf{D}_p(C_\alpha^+) \tag{12.21}$$

---

[5] The corresponding matrix representative for reflection across the plane specified by σ_α is
$\begin{pmatrix} \cos 2\alpha & \sin 2\alpha \\ \sin 2\alpha & -\cos 2\alpha \end{pmatrix}$.

Care must be taken in noting that the *same* operation $\hat{R}$ may have a different effect when acting on a *different* basis. The transformation under C_α^+ of the d-functions in the xy-plane:

$$d_{x^2-y^2} \propto \sin^2\theta\cos 2\varphi, \quad d_{xy} \propto \sin^2\theta\sin 2\varphi \tag{12.22}$$

gives in fact

$$C_\alpha^+(d_{x^2-y^2}d_{xy}) = (C_\alpha^+ d_{x^2-y^2} C_\alpha^+ d_{xy}) = (d_{x^2-y^2}d_{xy})\begin{pmatrix} \cos 2\alpha & \sin 2\alpha \\ -\sin 2\alpha & \cos 2\alpha \end{pmatrix}$$
$$= (d_{x^2-y^2}d_{xy})D_d(C_\alpha^+)$$

$$\tag{12.23}$$

Therefore, it is always necessary to specify the basis which the matrix representative refers to. These and other matrix representatives of different symmetry operations may be found elsewhere (Magnasco, 2007).

As a last point on symmetry, it must be recalled that

If $RS = T$ in coordinate space, then $\hat{R}\hat{S} = \hat{T}$ in function space

and $D(R)D(S) = D(T)$ in matrix space $\tag{12.24}$

12.2   GROUP THEORETICAL METHODS

Let us now briefly introduce the axioms defining the concept of a *group*.

An abstract group $G\{G_1, G_2, \ldots, G_h\}$ of order h is given by a *closed* set of h elements satisfying the following properties:

(i)   There is a *composition* law (usually, but not necessarily, the multiplication law) such that, for G_r and G_s belonging to G, $G_rG_s = G_t$ still belongs to G (we then say that a group is a set *closed* with respect to symbolic multiplication).

(ii)   The composition law is *associative*: $(G_rG_s)G_t = G_r(G_sG_t)$.

(iii)   There is an *identity* (or neutral) element G_μ, such that $G_rG_\mu = G_\mu G_r = G_r$.

(iv)   Each element has an *inverse* G_r^{-1}, such that $G_rG_r^{-1} = G_r^{-1}G_r = G_\mu$.

Table 12.1 The molecular point group C_{2v}

C_{2v}	I	C_2	σ_v	σ'_v
A_1	1	1	1	1
A_2	1	1	-1	-1
B_1	1	-1	-1	1
B_2	1	-1	1	-1

(v) In general, the commutative law does not hold for the symbolic multiplication; namely: $G_rG_s \neq G_sG_r$. If $G_rG_s = G_sG_r$, then the group is said to be Abelian (a *commutative* group).

Almost all textbooks on group theory contain tables of molecular point groups, their relation to the chemical structure of some polyatomic molecules being given for instance in Eyring *et al.* (1944). We give below two examples of molecular point groups and their multiplication tables (Tables 12.1–12.4): the Abelian group C_{2v} of the H_2O molecule and the group C_{3v} of NH_3. A point group of order h has h symmetry operations characterizing it. So, C_{2v} is a group of order 4, while C_{3v} is a group of order 6.

Tables 12.1 and 12.3 are called *character tables* of the point groups. In the first column of both tables, below the denomination of the point group, are the symbols denoting the symmetry-defined types or, in the

Table 12.2 The multiplication table of the point group C_{2v}

C_{2v}	I	C_2	σ_v	σ'_v
I	I	C_2	σ_v	σ'_v
C_2	C_2	I	σ'_v	σ_v
σ_v	σ_v	σ'_v	I	C_2
σ'_v	σ'_v	σ_v	C_2	I

Table 12.3 The molecular point group C_{3v}

C_{3v}	I	C_3^+	C_3^-	σ_1	σ_2	σ_3
A_1	1	1	1	1	1	1
A_2	1	1	1	-1	-1	-1
E	$\begin{pmatrix} 1 & 0 \\ 0 & 1 \end{pmatrix}$	$\begin{pmatrix} \bar{c} & s \\ \bar{s} & \bar{c} \end{pmatrix}$	$\begin{pmatrix} \bar{c} & \bar{s} \\ s & \bar{c} \end{pmatrix}$	$\begin{pmatrix} 1 & 0 \\ 0 & \bar{1} \end{pmatrix}$	$\begin{pmatrix} \bar{c} & \bar{s} \\ \bar{s} & c \end{pmatrix}$	$\begin{pmatrix} \bar{c} & s \\ s & c \end{pmatrix}$

Table 12.4   The multiplication table of the point group C_{3v}

C_{3v}	I	C_3^+	C_3^-	σ_1	σ_2	σ_3
I	I	C_3^+	C_3^-	σ_1	σ_2	σ_3
C_3^+	C_3^+	C_3^-	I	σ_2	σ_3	σ_1
C_3^-	C_3^-	I	C_3^+	σ_3	σ_1	σ_2
σ_1	σ_1	σ_3	σ_2	I	C_3^-	C_3^+
σ_2	σ_2	σ_1	σ_3	C_3^+	I	C_3^-
σ_3	σ_3	σ_2	σ_1	C_3^-	C_3^+	I

language of group theory, the *irreducible representations* (in short, irreps) of that group. For each irrep, symmetry operations are given in the form of matrices like (12.21) or (12.23), whose order equals the dimensionality of the irrep.[6] The *characters* are simply the trace of such matrices, and are the only numbers given in textbooks. While C_{2v} has only one-dimensional irreps, the higher symmetry C_{3v} group has two one-dimensional irreps (A_1 and A_2) and one two-dimensional irrep (E).[7]

Tables 12.2 and 12.4 give the multiplication tables of the two groups. In constructing a multiplication table, we recall that $R_k = R_i R_j$ is the result of the intersection of the column headed by R_i and the row headed by R_j, and that the operation on the right must be done first. It is seen that each symmetry operation occurs once in each row.

We now give a few further definitions and fundamental theorems.

12.2.1   Isomorphism

Two groups G and G' of the same order h are *isomorphic* if (i) there is a one-to-one correspondence between each element G_r of G and G'_r of $G'(r = 1, 2, \ldots, h)$ and (ii) the symbolic multiplication rule is preserved, namely, if $G_r G_s = G_t$ in $G \Longleftrightarrow G'_r G'_s = G'_t$ in G'. If only (ii) is true then the groups are said to be homomorphic.

12.2.2   Conjugation and Classes

Any two elements A and B of a group G are said to be *conjugate* if they are related by a similarity transformation with one other element X of the

---

[6] In C_{3v}, $c = 1/2$, $s = \sqrt{3}/2$.

[7] Cubic groups have also three-dimensional irreps (T).

group, namely: $A = X^{-1}BX$. The set of all conjugate elements defines a *class*. Conjugate operations are always of the same type (rotations with rotations, reflections with reflections, etc.). The number of classes equals the number of irreps.

12.2.3 Representations and Characters

Let $G\{G_1, G_2, \ldots, G_h\}$ be a group of h elements and $\{D(G_1), D(G_2), \ldots, D(G_h)\}$ a group of matrices isomorphic to G. We then say that the group of matrices gives a *representation* (German, *darstellung*, hence D) of the abstract group. If we have a representation of a group in the form of a group of matrices, then we also have an infinite number of representations. In fact, we can always subject all matrices of a given representation to a similarity transformation, thereby obtaining a *new* representation, and so on, the multiplication rule being preserved during the similarity transformation. If, by applying a similarity transformation with a unitary matrix U to a representation of a group G in the form of a group of matrices, we obtain a new representation whose matrices have a block-diagonal form, we say that the representation has been *reduced*.

The set of functions that are needed to find a (generally, *reducible*) representation Γ forms a *basis* for the representation. The functions forming a basis for the *irreducible representations* (irreps) of a symmetry group are said to be *symmetry-adapted functions*, and transform in the simplest and characteristic way under the symmetry operations of the group. It is of basic importance in quantum chemistry to find such symmetry-adapted functions, starting from a given basis set through use of suitable projection operators, as we shall see.

As already said before, the *character* is the trace of the matrix representative of the symmetry operator $\hat{R}$ and is denoted by $\chi(R)$. The characters have the following properties:

1. They are invariant under any transformation of the basis.
2. They are the same for all symmetry operations belonging to the same class.
3. The condition for two representations to be equivalent is that they have the same characters.

12.2.4 Three Theorems on Irreducible Representations

1. The necessary and sufficient condition that a representation Γ be irreducible is that the sum over all operations $\hat{R}$ of the group of the

squares of the moduli of the characters be equal to the order h of the group:

$$\sum_R \chi^*(R)\chi(R) = \sum_R |\chi(R)|^2 = h \qquad (12.25)$$

2. Given any two irreducible representations Γ^i and Γ^j of a group, we have the *orthogonality theorem* for the characters:

$$\sum_R \chi^i(R)^* \chi^j(R) = h\delta_{ij} \qquad (12.26)$$

3. This theorem is a particular case of the more general *orthogonality theorem* for the components of the representative matrices of the h elements of the group:

$$\sum_R D^i(R)^*_{mn} D^j(R)_{m'n'} = \frac{h}{\ell_i} \delta_{ij} \delta_{mm'} \delta_{nn'} \qquad (12.27)$$

where ℓ_i is the dimensionality of the ith irrep.

12.2.5   Number of Irreps in a Reducible Representation

The number of times a_j a given irrep Γ^j occurs in the reducible representation Γ follows from the orthogonality theorem and is given by

$$a_j = \frac{1}{h} \sum_R \chi^j(R)^* \chi^\Gamma(R) \qquad (12.28)$$

12.2.6   Construction of Symmetry-adapted Functions

Symmetry-adapted functions transforming as the λ-component of the jth irreducible representation Γ^j are obtained by use of the projector

$$\hat{P}^j_{\lambda\lambda} = \frac{\ell_j}{h} \sum_R D^j(R)^*_{\lambda\lambda} \hat{R} \qquad (12.29)$$

which requires the complete knowledge of matrices $\mathbf{D}^j(R)$ for each operation R of the group. This projector is needed in the case of multi-dimensional irreps, but for one-dimensional irreps it is sufficient to use the simpler projector involving the characters

$$\hat{P}^j \propto \sum_R \chi^j(R)^* \hat{R} \qquad (12.30)$$

12.3   ILLUSTRATIVE EXAMPLES

Use of symmetry-adapted functions is particularly useful for predicting which matrix elements of an operator are zero (selection rules, mostly in spectroscopy), even without doing any effective calculation, and in the factorization of the secular equations arising either in MO, CI, or VB calculations. We shall illustrate below these factorizations for the minimal basis set Hückel calculations for the H_2O (C_{2v}) and NH_3 (C_{3v}) molecules.

12.3.1   Use of Symmetry in Ground-state H_2O ($^1A_1$)

The H_2O molecule is chosen to lie in the yz-plane, with the O atom at the origin of a right-handed coordinate system having z as binary symmetry axis (Figure 12.2). The nuclear symmetry implies two symmetry planes (yz

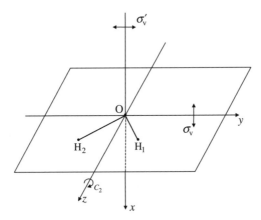

Figure 12.2   Elements of C_{2v} symmetry in H_2O

and zx), their intersection determining the twofold symmetry axis C_2. The molecular point group is hence C_{2v}, whose character table was given in Table 12.1. The minimal set for the MO calculation is given, in an obvious brief notation, by the row matrix of the $m = 7$ STOs:

$$\chi = (k\, s\, z\, x\, y\, h_1\, h_2) \tag{12.31}$$

which must be combined linearly to give seven MOs, the first $n = 5$ being doubly occupied by electrons with opposite spin, so accomodating the $N = 2n = 10$ electrons of the molecule in its totally symmetric singlet ${}^1A_1$ ground state. The LCAO coefficients of the Roothaan equations are obtained by the Ritz method through the iterative solution of a (7×7) secular equation, but for our illustrative purposes it will be sufficient to consider the simple Hückel approximation in the same basis. The construction of the symmetry-adapted basis for the calculation can be done at once by simple inspection, since it is immediately evident that the functions belonging to the different irreps of the point group C_{2v} are

$$k, s, z, h_z = \frac{1}{\sqrt{2}}(h_1 + h_2) \Rightarrow A_1 \tag{12.32}$$

$$x \Rightarrow B_1 \tag{12.33}$$

$$y, h_y = \frac{1}{\sqrt{2}}(h_1 - h_2) \Rightarrow B_2 \tag{12.34}$$

Since functions of different symmetry cannot mix under the totally symmetric Hückel operator H, the (7×7) Hückel representative $\mathbf{H}$ over the minimum set will be factorized into three blocks of different symmetries: A_1 (4×4), B_1 (1×1) and B_2 (2×2).

Turning to the more formal group theoretical techniques, we give in Table 12.5 the dimensions of the representative matrices for the different operations R in the *reducible* representation Γ in the original basis, and those of the *irreducible* representations Γ^j corresponding to the symmetry-adapted functions of the last column. The latter are easily obtained by letting the projector (12.30) act on the original AO basis (12.31). It is left to the reader to verify in this case all the

Table 12.5 Reducible representation Γ and symmetry-adapted AOs for H_2O

C_{2v}	I	C_2	σ_v	σ'_v	Symmetry basis
A_1	1	1	1	1	$k, s, z, h_z = \frac{1}{\sqrt{2}}(h_1 + h_2)$
A_2	1	1	-1	-1	
B_1	1	-1	-1	1	x
B_2	1	-1	1	-1	$y, h_y = \frac{1}{\sqrt{2}}(h_1 - h_2)$
Γ	7	1	5	3	$\Gamma^{A_1} = 4, \quad \Gamma^{B_1} = 1, \quad \Gamma^{B_2} = 2$

properties and theorems introduced in the formal group theory of the preceding section.

Just to be clear, we give below the construction of the matrix representative $\mathbf{D}(C_2)$ for the *reducible* representation Γ in the original basis (12.31). We must first construct the *transformation table* of the basis functions (12.31) under the symmetry operations of C_{2v}. Using the active transformation, we obtain Table 12.6.

Following the recipe (12.9), we then immediately obtain

$$\mathbf{D}(C_2) = \begin{pmatrix} 1 & \cdot & \cdot & \cdot & \cdot & \cdot & \cdot \\ \cdot & 1 & \cdot & \cdot & \cdot & \cdot & \cdot \\ \cdot & \cdot & 1 & \cdot & \cdot & \cdot & \cdot \\ \cdot & \cdot & \cdot & \bar{1} & \cdot & \cdot & \cdot \\ \cdot & \cdot & \cdot & \cdot & \bar{1} & \cdot & \cdot \\ \cdot & \cdot & \cdot & \cdot & \cdot & \cdot & 1 \\ \cdot & \cdot & \cdot & \cdot & \cdot & 1 & \cdot \end{pmatrix}, \quad \mathrm{tr}\,\mathbf{D}(C_2) = 1 \qquad (12.35)$$

Table 12.6 Transformation table of the AO basis for H_2O under the operations of C_{2v}

$\hat{R}\chi$	I	C_2	σ_v	σ'_v
k	k	k	k	k
s	s	s	s	s
z	z	z	z	z
x	x	$-x$	$-x$	x
y	y	$-y$	y	$-y$
h_1	h_1	h_2	h_1	h_2
h_2	h_2	h_1	h_2	h_1

as reported in the last row of Table 12.5. We can proceed similarly for $D(I)$, $D(\sigma_v)$, and $D(\sigma'_v)$. Using the standard techniques of the matrix eigenvalue problem, it is left as an easy exercise for the reader to verify that the lower off-diagonal block of the representative matrices $D(C_2)$ and $D(\sigma_v)$ in the original basis set (12.31) are *diagonalized* in the symmetry-adapted basis $(h_z h_y)$, namely by transformation with the unitary matrix

$$
U = \begin{pmatrix} \dfrac{1}{\sqrt{2}} & \dfrac{1}{\sqrt{2}} \\[2ex] \dfrac{1}{\sqrt{2}} & -\dfrac{1}{\sqrt{2}} \end{pmatrix} \tag{12.36}
$$

12.3.2   Use of Symmetry in Ground-state NH_3 ($^1A_1$)

The three hydrogen atoms of the NH_3 molecule are chosen to lie in the xy-plane, numbered anticlockwise with H_1 on the positive x-axis, the origin of the right-handed coordinate system being taken at their intersection and the N atom along the symmetry axis with z positive (Figure 12.3). The nuclear symmetry now implies three symmetry planes at $120°$ ($\sigma_1, \sigma_2, \sigma_3$), their intersection determining the threefold symmetry axis C_3. The

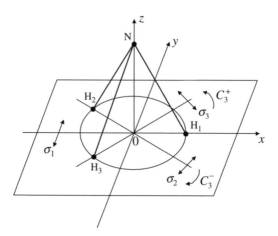

Figure 12.3   Elements of C_{3v} symmetry in NH_3

molecular point group is hence C_{3v}, whose character table was given in Table 12.3. The minimal basis set for the MO calculation is given, in an obvious brief notation, by the row matrix of the $m = 8$ STOs:

$$\chi = (k \, s \, z \, x \, y \, h_1 \, h_2 \, h_3) \qquad (12.37)$$

which must be combined linearly to give eight MOs, the first $n = 5$ being doubly occupied by electrons with opposite spin, so accommodating the $N = 2n = 10$ electrons of the molecule in its totally symmetric singlet $^1A_1$ ground state.

Even in this case, the construction of the symmetry-adapted basis for the calculation can be done at once by simple inspection, since it is immediately evident that the functions belonging to the different irreps of the point group C_{3v} are

$$k, s, z, h_z = \frac{1}{\sqrt{3}} (h_1 + h_2 + h_3) \Rightarrow A_1 \qquad (12.38)$$

$$\begin{cases} x, h_x = \dfrac{1}{\sqrt{6}} (2h_1 - h_2 - h_3) \\[2mm] y, h_y = \dfrac{1}{\sqrt{2}} (h_2 - h_3) \end{cases} \Rightarrow E \qquad (12.39)$$

the last being a doubly degenerate irrep whose basic vectors transform as $(x \, y)$.

Turning to the group theoretical techniques, following what was done before for H_2O, Table 12.7 gives the dimensions of the representative matrices for the different operations R in the *reducible* representation Γ in the original basis and those of the *irreducible* representations Γ^j corresponding to the symmetry-adapted functions of the last column. The latter are obtained by letting the full projector (12.29) act on the original AO basis (12.37). Table 12.8 is the transformation table of the original AO basis (recall that in the table $c = 1/2$, $s = \sqrt{3}/2$).

In the case of NH_3, because of the presence of the doubly degenerate irrep E, the simple projector (12.30) based on characters would yield nonorthogonal linearly dependent symmetry functions, which should then be Schmidt orthogonalized to give a linearly independent set. It is

Table 12.7 Reducible representation Γ and symmetry-adapted AOs for NH_3

C_{3v}	I	C_3^+	C_3^-	σ_1	σ_2	σ_3	Symmetry basis
A_1	1	1	1	1	1	1	$k,s,z,h_z=\frac{1}{\sqrt{3}}(h_1+h_2+h_3)$
A_2	1	1	1	-1	-1	-1	
E	$\begin{pmatrix}1&0\\0&1\end{pmatrix}$	$\begin{pmatrix}\bar c&s\\\bar s&\bar c\end{pmatrix}$	$\begin{pmatrix}\bar c&\bar s\\s&\bar c\end{pmatrix}$	$\begin{pmatrix}1&0\\0&\bar1\end{pmatrix}$	$\begin{pmatrix}\bar c&\bar s\\\bar s&c\end{pmatrix}$	$\begin{pmatrix}\bar c&s\\s&c\end{pmatrix}$	$\begin{cases}x,h_x=\frac{1}{\sqrt6}(2h_1-h_2-h_3)\\[2mm]y,h_y=\frac{1}{\sqrt2}(h_1-h_2)\end{cases}$
Γ	8	2	2	4	4	4	$\Gamma^{A_1}=4,\ \Gamma^{E_x}=2,\ \Gamma^{E_y}=2$

convenient, therefore, to use at once the full projector (12.29), as we shall show in detail below.

$$A_1:\quad \hat P^{A_1}k=k\quad \hat P^{A_1}s=s\quad \hat P^{A_1}z=z$$

$$\hat P^{A_1}x=\frac16(x-cx-sy-cx+sy+x-cx-sy-cx+sy)=0$$

$$\hat P^{A_1}y=\frac16(y+sx-cy-sx-cy-y-sx+cy+sx+cy)=0$$

$$\hat P^{A_1}h_1=\frac16(2h_1+2h_2+2h_3)=\frac13(h_1+h_2+h_3)\Rightarrow\frac{1}{\sqrt3}(h_1+h_2+h_3)$$

Table 12.8 Transformation table of the AO basis for NH_3 under the operations of C_{3v}

$\hat R\chi$	I	C_3^+	C_3^-	σ_1	σ_2	σ_3
k	k	k	k	k	k	k
s	s	s	s	s	s	s
z	z	z	z	z	z	z
x	x	$\bar c x+\bar s y$	$\bar c x+sy$	x	$\bar c x+\bar s y$	$\bar c x+sy$
y	y	$sx+\bar c y$	$\bar s x+\bar c y$	$-y$	$\bar s x+cy$	$sx+cy$
h_1	h_1	h_3	h_2	h_1	h_3	h_2
h_2	h_2	h_1	h_3	h_3	h_2	h_1
h_3	h_3	h_2	h_1	h_2	h_1	h_3

$$A_2: \quad \hat{P}^{A_2}k = \frac{1}{6}(k+k+k-k-k-k) = 0 \quad \hat{P}^{A_2}s = \hat{P}^{A_2}z = 0$$

$$\hat{P}^{A_2}x = \frac{1}{6}(x-cx-sy-cx+sy-x+cx+sy+cx-sy) = 0$$

$$\hat{P}^{A_2}y = \frac{1}{6}(y+sx-cy-sx-cy+y+sx-cy-sx-cy) = 0$$

$$\hat{P}^{A_2}h_1 = \frac{1}{6}(h_1+h_3+h_2-h_1-h_3-h_2) = 0$$

$$E: \quad \hat{P}^{E}_{xx}k = \frac{2}{6}(k-ck-ck+k-ck-ck) = 0 \quad \hat{P}^{E}_{xx}s = \hat{P}^{E}_{xx}z = 0$$

$$\begin{cases} \hat{P}^{E}_{xx}x = \frac{2}{6}[x-c(-cx-sy)-c(-cx+sy)+x-c(-cx-sy)-c(-cx+sy)] \\[2mm] \qquad = \frac{1}{3}(2x+c^2x+c^2x+c^2x+c^2x) = x \end{cases}$$

$$\begin{cases} \hat{P}^{E}_{xx}y = \frac{2}{6}[y-c(sx-cy)-c(-sx-cy)-y-c(-sx+cy)-c(sx+cy)] \\[2mm] \qquad = \frac{1}{3}(y-csx+c^2y+csx+c^2y-y+csx-c^2y-csx-c^2y) = 0 \end{cases}$$

$$\begin{cases} \hat{P}^{E}_{xx}h_1 = \frac{2}{6}(h_1-ch_3-ch_2+h_1-ch_3-ch_2) = \frac{1}{3}(2h_1-h_2-h_3) \\[2mm] \qquad \Rightarrow \frac{1}{\sqrt{6}}(2h_1-h_2-h_3) \end{cases}$$

$$\hat{P}^{E}_{yy}k = \frac{2}{6}(k-ck-ck-k+ck+ck) = 0 \quad \hat{P}^{E}_{yy}s = \hat{P}^{E}_{yy}z = 0$$

$$\begin{cases} \hat{P}^{E}_{yy}x = \frac{2}{6}[x-c(-cx-sy)-c(-cx+sy)-x+c(-cx-sy)+c(-cx+sy)] \\[2mm] \qquad = \frac{1}{3}(x+c^2x+csy+c^2x-csy-x-c^2x-csy-c^2x+csy) = 0 \end{cases}$$

$$\begin{cases} \hat{P}_{yy}^E y = \dfrac{2}{6}[y - c(sx-cy) - c(-sx-cy) + y + c(-sx+cy) + c(sx+cy)] \\[2mm] \quad = \dfrac{1}{3}(y - csx + c^2y + csx + c^2y + y - csx + c^2y + csx + c^2y) = y \end{cases}$$

$$\hat{P}_{yy}^E h_1 = \dfrac{2}{6}(h_1 - ch_3 - ch_2 - h_1 + ch_3 + ch_2) = 0$$

$$\begin{cases} \hat{P}_{yy}^E h_2 = \dfrac{2}{6}(h_2 - ch_1 - ch_3 - h_3 + ch_2 + ch_1) = \dfrac{1}{3}\left(\dfrac{3}{2}h_2 - \dfrac{3}{2}h_3\right) \\[2mm] \Rightarrow \dfrac{1}{\sqrt{2}}(h_2 - h_3) \end{cases}$$

The reducible representation Γ of Table 12.7 splits into the following irreps:

$$\Gamma = 4A_1 + 2E \tag{12.40}$$

in accord with the result of Equation 12.28.

Therefore, the (8 × 8) secular equation in the original AO basis factorizes into a (4 × 4) block of symmetry A_1 and two (2 × 2) blocks of symmetry E, belonging to the symmetries E_x and E_y (Figure 12.4) which are mutually orthogonal and not interacting.

We end by recalling that MOs resulting from all LCAO methods are classified according to the irreducible representations to which they

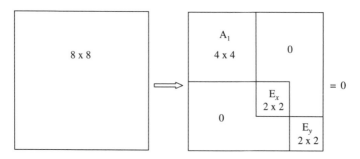

Figure 12.4 Factorization under C_{3v} of the minimum basis set MO secular equation for NH_3

belong, using lower case letters preceded by a principal quantum number n specifying the order of the orbital energies (Roothaan, 1951b). In this way, the electron configurations of ground and excited states of molecules are given much in the same way as those familiar for atoms. In our case of the $N = 10$-electron molecules H_2O and NH_3, the electron configurations of their totally symmetric singlet $^1A_1$ ground states are given by

$$H_2O\ (C_{2v}, {}^1A_1):\ 1a_1^2\ 2a_1^2\ 1b_2^2\ 3a_1^2\ 1b_1^2 \tag{12.41}$$

$$NH_3\ (C_{3v}, {}^1A_1):\ 1a_1^2\ 2a_1^2\ 1e^4\ 3a_1^2 \tag{12.42}$$

References

Abbott, P.C. (2007) Asymptotic expansion of the Keesom integral. *J. Phys. A: Math. Gen.*, **40**, 8599–8600.

Abramowitz, M. and Stegun, I.A. (1965) *Handbook of Mathematical Functions*, Dover, New York.

Aitken, A.C. (1958) *Determinants and Matrices*, Oliver and Boyd, Edinburgh.

Atkin, R.H. (1959) *Classical Dynamics*, Heinemann, London.

Atkins, P.W. and Friedman, R.S. (2007) *Molecular Quantum Mechanics*, 4th edn, Oxford University Press, Oxford.

Babb, J.F. (1994) Effective oscillator strengths and transition energies for the hydrogen molecular ion. *Mol. Phys.*, **81**, 17–29.

Bacon, G.E. (1969) The applications of neutron diffraction. *Endeavour*, **25**, 129–135.

Ballinger, R.A. (1959) The electronic structure of hydrogen fluoride. *Mol. Phys.*, **2**, 139–157.

Bartlett, R.J., Shavitt, I., and Purvis, G.D. III (1979) The quartic force field of H_2O determined by many-body methods that include quadruple excitation effects. *J. Chem. Phys.*, **71**, 281–291.

Battezzati, M. and Magnasco, V. (2004) Asymptotic evaluation of the Keesom integral. *J. Phys. A: Math. Gen.*, **37**, 9677–9684.

Bishop, D.M. and Cheung, L.M. (1978) Moment functions (including static dipole polarisabilities) and radiative corrections for H_2^+. *J. Phys. B: Atom. Mol. Phys.*, **11**, 3133–3144.

Bishop, D.M., Pipin, J., and Cybulski, S.M. (1991) Theoretical investigation of the nonlinear optical properties of H_2 and D_2: extended basis set. *Phys. Rev. A*, **43**, 4845–4853.

Born, M. (1962) *Atomic Physics*, 7th edn, Blackie, London.

Born, M. and Oppenheimer, R. (1927) Zur Quantentheorie der Molekeln. *Ann. Phys. Leipzig*, **84**, 457–484.

Boys, S.F. (1950) Electronic wave functions. I. A general method of calculation for stationary states of any molecular system. *Proc. R. Soc. London Ser. A*, **200**, 542–554.

268   REFERENCES

Briggs, M.P., Murrell, J.N., and Salahub, D.R. (1971) Variational calculations of the static polarizability of H_2 and of the dipole–dipole contribution to the dispersion energy between two hydrogen molecules. *Mol. Phys.*, **22**, 907–913.

Brink, D.M. and Satchler, G.R. (1993) *Angular Momentum*, 3rd edn, Clarendon Press, Oxford.

Buckingham, A.D. (1982) Closing remarks. *Faraday Discuss. Chem. Soc.*, **73**, 421–423.

Cade, P.E. and Huo, W.H. (1967) Electronic structure of diatomic molecules. VI.A. Hartree–Fock wavefunctions and energy quantities for the ground states of the first-row hydrides, AH. *J. Chem. Phys.*, **47**, 614–648.

Casimir, H.B.G. and Polder, D. (1948) The influence of retardation on the London–van der Waals forces. *Phys. Rev.*, **73**, 360–372.

Cencek, W. and Rychlewski, J. (1995) Many-electron explicitly correlated Gaussian functions. II. Ground state of the helium molecular ion He_2^+. *J. Chem. Phys.*, **102**, 2533–2538.

Clementi, E. (1962) SCF-MO wave functions for the hydrogen fluoride molecule. *J. Chem. Phys.*, **36**, 33–44.

Clementi, E. and Corongiu, G. (1982) *Geometrical Basis Set For Molecular Computations*, IBM IS&TG, pp. 1–182.

Clementi, E. and Roetti, C. (1974) Roothaan–Hartree–Fock atomic wave functions. *Atom. Data Nucl. Data Tables*, **14**, 177–478.

Cooper, D.L., Gerratt, J., and Raimondi, M. (1986) The electronic structure of the benzene molecule. *Nature*, **323**, 699–701.

Coulson, C.A. (1958) *Electricity*, 5th edn, Oliver and Boyd, Edinburgh.

Coulson, C.A. (1961) *Valence*, 2nd edn, Oxford University Press, Oxford.

Davies, D.W. (1967) *The Theory of the Electric and Magnetic Properties of Molecules*, John Wiley & Sons, London.

Dirac, P.A.M. (1929) Quantum mechanics of many-electron systems. *Proc. R. Soc. London Ser. A*, **123**, 714–733.

Dixon, R.N. (1965) *Spectroscopy and Structure*, Methuen, London.

Dunning, T.H. Jr (1989) Gaussian basis sets for use in correlated molecular calculations. I. The atoms boron through neon and hydrogen. *J. Chem. Phys.*, **90**, 1007–1023.

Eckart, C. (1930) The theory and calculation of screening constants. *Phys. Rev.*, **36**, 878–892.

Eyring, H., Walter, J., and Kimball, G.E. (1944) *Quantum Chemistry*, John Wiley & Sons, Inc., New York.

Figari, G. and Magnasco, V. (2003) On the interpolation of frequency-dependent polar-izabilities through a readily integrable expression. *Chem. Phys. Lett.*, **374**, 527–533.

Figari, G., Rui, M., Costa, C., and Magnasco, V. (2007) C_6 dispersion coefficients from reduced dipole pseudospectra. *Theor. Chem. Acc.*, **118**, 107–112.

Fock, V. (1930) Näherungsmethode zur Lösung des quantenmechanischen Mehrkörper-problems. *Z. Phys*, **61**, 126–148.

Frankowski, K. and Pekeris, C.L. (1966) Logarithmic terms in the wave functions of the ground state of two-electron atoms. *Phys. Rev.*, **146**, 46–49.

Hall, G.G. (1951) The molecular orbital theory of chemical valency. VIII. A method of calculating ionization potentials. *Proc. R. Soc. London Ser. A*, **205**, 541–552.

Hartree, D.R. (1928a) The wave mechanics of an atom with a non-Coulomb central field. Part I. Theory and methods. *Proc. Cambridge Phil. Soc.*, **24**, 89–110.

Hartree, D.R. (1928b) The wave mechanics of an atom with a non-Coulomb central field. Part II. Some results and discussion. *Proc. Cambridge Phil. Soc.*, **24**, 111–132.

Heisenberg, W. (1930) *Die Physikalische Prinzipien der Quantentheorie*, Hirzel, Leipzig.

Heitler, W. und London, F. (1927) Wechselwirkung neutraler Atome und homöopolare Bindung nach der Quantenmechanik. *Z. Phys.*, **44**, 455–472.

Hobson, E.W. (1965) *The Theory of Spherical and Ellipsoidal Harmonics*, 2nd edn, Chelsea, New York.

Hohenberg, P. and Kohn, W. (1964) Inhomogeneous electron gas. *Phys. Rev.*, **136**, B864–B871.

Hohn, F.E. (1964) *Elementary Matrix Algebra*, 2nd edn, MacMillan, New York.

Huber, K.P. and Herzberg, G. (1979) *Molecular Spectra and Molecular Structure: IV. Constants of Diatomic Molecules*, Van Nostrand-Reinhold, New York.

Hückel, E. (1931) Quantentheoretische Beiträge zum Benzolproblem. I. Die Elektronen-konfiguration des Benzols und verwandter Verbindungen. *Z. Phys.*, **70**, 204–286.

Ince, E.L. (1963) *Integration of Ordinary Differential Equations*, 7th edn, Oliver and Boyd, Edinburgh.

Karplus, M. and Kolker, H.J. (1961) Magnetic interactions in lithium hydride. *J. Chem. Phys.*, **35**, 2235–2236.

Karplus, M. and Kolker, H.J. (1963) Magnetic susceptibility of diatomic molecules. *J. Chem. Phys.*, **38**, 1263–1275.

Kato, T. (1957) On the eigenfunctions of many-particle systems in quantum mechanics. *Commun. Pure Appl. Math.*, **10**, 151–177.

Keesom, W.H. (1921) Die van der Waalsschen Kohäsionkräfte. *Phys. Z.*, **22**, 129–141.

Klopper, W. and Kutzelnigg, W. (1991) Wave functions with terms linear in the inter-electronic coordinates to take care of the correlation cusp. III. Second-order Møller–Plesset (MP2-R12) calculations on molecules of first row atoms. *J. Chem. Phys.*, **94**, 2020–2031.

Koga, T. and Matsumoto, S. (1985) An exact solution of the Van der Waals interaction between two ground-state hydrogen atoms. *J. Chem. Phys.*, **82**, 5127–5132.

Kohn, W. and Sham, L.J. (1965) Self-consistent equations including exchange and correlation effects. *Phys. Rev.*, **140**, A1133–A1138.

Kutzelnigg, W. (1985) r_{12}-Dependent terms in the wave function as closed sums of partial wave amplitudes for large l. *Theor. Chim. Acta Berlin*, **68**, 445–469.

Lazzeretti, P. and Zanasi, R. (1981) On the theoretical determination of molecular first hyperpolarizability. *J. Chem. Phys.*, **74**, 5216–5224.

Lennard-Jones, J.E. (1931) Wave functions of many-electron atoms. *Proc. Cambridge Phil. Soc.*, **27**, 469–480.

Lennard-Jones, J.E. (1937) The electronic structure of some polyenes and aromatic molecules. I. The nature of the links by the method of molecular orbitals. *Proc. R. Soc. London Ser. A*, **158**, 280–296.

Liu, B. and McLean, A.D. (1973) Accurate calculation of the attractive interaction of two ground state helium atoms. *J. Chem. Phys.*, **59**, 4557–4558.

Longuet-Higgins, H.C. (1956) The electronic states of composite systems. *Proc. R. Soc. London Ser. A*, **235**, 537–543.

Longuet-Higgins, H.C. (1961) Some recent developments in the theory of molecular energy levels. *Adv. Spectr.*, **2**, 429–472.

Löwdin, P.O. (1955) Quantum theory of many-particle systems. I. Physical interpretations by means of density matrices, natural spin-orbitals, and convergence problems in the method of configurational interaction. *Phys. Rev.*, **97**, 1474–1489.

MacDonald, J.K.L. (1933) Successive approximations by the Rayleigh–Ritz variation method. *Phys. Rev.*, **43**, 830–833.

MacRobert, T.M. (1947) *Spherical Harmonics*, 2nd edn, Methuen, London.

Magnasco, V. (1978) A generalised London formula for dispersion coefficients. *Chem. Phys. Lett.*, **57**, 573–578.

Magnasco, V. (2002) On the α and β parameters in Hückel theory including overlap for simple σ molecular systems. *Chem. Phys. Lett.*, **363**, 544–549.

Magnasco, V. (2003) A model for the heteropolar bond. *Chem. Phys. Lett.*, **380**, 397–403.

Magnasco, V. (2004a) A model for the chemical bond. *J. Chem. Edu.*, **81**, 427–435.

Magnasco, V. (2004b) A model for the van der Waals bond. *Chem. Phys. Lett.*, **387**, 332–338.

Magnasco, V. (2005) On the principle of maximum overlap in molecular orbital theory. *Chem. Phys. Lett.*, **407**, 213–216.

Magnasco, V. (2007) *Elementary Methods of Molecular Quantum Mechanics*, Elsevier, Amsterdam.

Magnasco, V. (2008) Orbital exponent optimization in elementary VB calculations of the chemical bond in the ground state of simple molecular systems. *J. Chem. Edu.*, **85**, 1686–1691.

Magnasco, V., Battezzati, M., Rapallo, A., and Costa, C. (2006) Keesom coefficients in gases. *Chem. Phys. Lett.*, **428**, 231–235.

Magnasco, V. and Costa, C. (2005) On the principle of maximum overlap in valence bond theory. *Chem. Phys. Lett.*, **403**, 303–307.

Magnasco, V., Costa, C., and Figari, G. (1989) Long-range second-order interactions and the shape of the He–HF and Ne–HF complexes. *Chem. Phys. Lett.*, **156**, 585–591.

Magnasco, V., Costa, C., and Figari, G. (1990a) A model for the elementary prediction of the angular geometries of van der Waals dimers. *J. Mol. Struct. Theochem.*, **204**, 229–246.

Magnasco, V., Costa, C., and Figari, G. (1990b) Long-range coefficients for molecular interactions. II. *J. Mol. Struct. Theochem.*, **206**, 235–252.

Magnasco, V., Figari, G., and Costa, C. (1988) Long-range coefficients for molecular interactions. *J. Mol. Struct. Theochem.*, **164**, 49–66.

Magnasco, V. and Ottonelli, M. (1996a) Accurate evaluation of C_6 dispersion coefficients for $(H_2)_2$. *Chem. Phys. Lett.*, **248**, 82–88.

Magnasco, V. and Ottonelli, M. (1996b) Dipole polarizabilities and C_6 dispersion coefficients for small atomic and molecular systems. *Chem. Phys. Lett.*, **259**, 307–312.

Magnasco, V. and Ottonelli, M. (1999a) Long-range dispersion coefficients from a generalization of the London formula. *Trends in Chem. Phys.*, **7**, 215–232.

Magnasco, V. and Ottonelli, M. (1999b) Dipole polarizability pseudospectra and C_6 dispersion coefficients for $H_2{}^+$–$H_2{}^+$. *J. Mol. Struct. Theochem*, **469**, 31–40.

Magnasco, V., Ottonelli, M., Figari, G. *et al.* (1998) Polarizability pseudospectra and dispersion coefficients for H(1s)–H(1s). *Mol. Phys.*, **94**, 905–908.

Margenau, H. (1961) Fundamental principles of quantum mechanics, in *Quantum Theory*, vol. I (ed. D.R. Bates), Academic Press, New York.

Margenau, H. and Murphy, G.M. (1956) *The Mathematics of Physics and Chemistry*, Van Nostrand, Princeton.

McLean, A.D. and Yoshimine, M. (1967a) Theory of molecular polarizabilities. *J. Chem. Phys.*, **47**, 1927–1935.

McLean, A.D. and Yoshimine, M. (1967b) Computed ground-state properties of FH and CH. *J. Chem. Phys.*, **47**, 3256–3262.

McWeeny, R. (1950) Gaussian approximations to wave functions. *Nature*, **166**, 21–22.

McWeeny, R. (1960) Some recent advances in density matrix theory. *Rev. Mod. Phys.*, **32**, 335–369.

McWeeny, R. (1979) *Coulson's Valence*, 3rd edn, Oxford University Press, Oxford.

McWeeny, R. (1989) *Methods of Molecular Quantum Mechanics*, 2nd edn, Academic Press, London.

Meyer, W. (1976) Dynamic multipole polarizabilities of H_2 and He and long-range interaction coefficients for H_2–H_2, H_2–He and He–He. *Chem. Phys.*, **17**, 27–33.

Mohr, P.J. and Taylor, B.N. (2003) The fundamental physical constants. *Phys. Today*, **56**, BG6–BG13.

Møller, C. and Plesset, M.S. (1934) Note on an approximation treatment for many-electron systems. *Phys. Rev.*, **46**, 618–622.

Moore, C.E. (1971) *Atomic Energy Levels as Derived From the Analyses of Optical Spectra*, vol. I, US National Bureau of Standards Circular 467, Government Printing Office, Washington, DC, pp. 4–5.

Mulder, J.J.C. (1966) On the number of configurations in an N-electron system. *Mol. Phys.*, **10**, 479–488.

Mulliken, R.S. (1955) Electronic population analysis on LCAO-MO molecular wave functions. I. *J. Chem. Phys.*, **23**, 1833–1840.

Murrell, J.N., Kettle, S.F.A., and Tedder, J.M. (1985) *The Chemical Bond*, 2nd edn, John Wiley & Sons, Inc., New York.

Noga, J. and Kutzelnigg, W. (1994) Coupled cluster theory that takes care of the correlation cusp by inclusion of linear terms in the interelectronic coordinates. *J. Chem. Phys.*, **101**, 7738–7762.

Noga, J., Valiron, P., and Klopper, W. (2001) The accuracy of atomization energies from explicitly correlated coupled-cluster calculations. *J. Chem. Phys.*, **115**, 2022–2032.

Pauli, W. Jr (1926) Über das Wasserstoffspektrum vom Standpunkt der neuen Quantenmechanik. *Z. Phys.*, **36**, 336–363.

Pauling, L. (1933) The calculation of matrix elements for Lewis electronic structures of molecules. *J. Chem. Phys.*, **1**, 280–283.

Peek, J.M. (1965) Eigenparameters for the $1s\sigma_g$ and $2p\sigma_u$ orbitals of $H_2{}^+$. *J. Chem. Phys.*, **43**, 3004–3006.

Pekeris, C.L. (1958) Ground state of two-electron atoms. *Phys. Rev.*, **112**, 1649–1658.

Pekeris, C.L. (1959) $1\,{}^1S$ and $2\,{}^3S$ states of helium. *Phys. Rev.*, **115**, 1216–1221.

Ritz, W. (1909) Über eine neue Methode zur Lösung gewisser Variationsprobleme der mathematischen Physik. *J. Reine Angew. Math.*, **135**, 1–61.

Roos, B. (1972) A new method for large-scale CI calculations. *Chem. Phys. Lett.*, **15**, 153–159.

Roothaan, C.C.J. (1951a) A study of two-center integrals useful in calculations on molecular structure. I. *J. Chem. Phys.*, **19**, 1445–1458.

Roothaan, C.C.J. (1951b) New developments in molecular orbital theory. *Rev. Mod. Phys.*, **23**, 69–89.

Rosenberg, B.J., Ermler, W.C., and Shavitt, I. (1976) Ab initio SCF and CI studies on the ground state of the water molecule. II. Potential energy and property surfaces. *J. Chem. Phys.*, **65**, 4072–4080.

Rosenberg, B.J. and Shavitt, I. (1975) Ab initio SCF and CI studies on the ground state of the water molecule. I. Comparison of CGTO and STO basis sets near the Hartree–Fock limit. *J. Chem. Phys.*, **63**, 2162–2174.

Ruedenberg, K. (1962) The physical nature of the chemical bond. *Rev. Mod. Phys.*, **34**, 326–376.

Rutherford, D.E. (1962) *Vector Methods*, 9th edn, Oliver and Boyd, Edinburgh.

Schroedinger, E. (1926a) Quantisierung als Eigenwertproblem (Erste Mitteilung). *Ann. Phys.*, **79**, 361–376.

Schroedinger, E. (1926b) Quantisierung als Eigenwertproblem (Zweite Mitteilung). *Ann. Phys.*, **79**, 489–527.

Schroedinger, E. (1926c) Quantisierung als Eigenwertproblem (Dritte Mitteilung). *Ann. Phys.*, **80**, 437–490.

Schroedinger, E. (1926d) Quantisierung als Eigenwertproblem (Vierte Mitteilung). *Ann. Phys.*, **81**, 109–139.

Schroedinger, E. (1926e) Über das Verhältnis der Heisenberg–Born–Jordanschen Quantenmechanik zu der meinen. *Ann. Phys.*, **79**, 734–756.

Slater, J.C. (1929) The theory of complex spectra. *Phys. Rev.*, **34**, 1293–1322.

Slater, J.C. (1930) Atomic shielding constants. *Phys. Rev.*, **36**, 57–64.

Slater, J.C. (1931) Molecular energy levels and valence bonds. *Phys. Rev.*, **38**, 1109–1144.

Stevens, R.M. and Lipscomb, W.N. (1964a) Perturbed Hartree–Fock calculations. II. Further results for diatomic lithium hydride. *J. Chem. Phys.*, **40**, 2238–2247.

Stevens, R.M. and Lipscomb, W.N. (1964b) Perturbed Hartree–Fock calculations. III. Polarizability and magnetic properties of the HF molecule. *J. Chem. Phys.*, **41**, 184–194.

Stevens, R.M. and Lipscomb, W.N. (1964c) Perturbed Hartree–Fock calculations. IV. Second-order properties of the fluorine molecule. *J. Chem. Phys.*, **41**, 3710–3716.

Stevens, R.M., Pitzer, R.M., and Lipscomb, W.N. (1963) Perturbed Hartree–Fock calculations. I. Magnetic susceptibility and shielding in the LiH molecule. *J. Chem. Phys.*, **38**, 550–560.

Strand, T.G. and Bonham, R.A. (1964) Analytical expressions for the Hartree–Fock potential of neutral atoms and for the corresponding scattering factors for X-rays and electrons. *J. Chem. Phys.*, **40**, 1686–1691.

Sundholm, D. (1985) Applications of fully numerical two-dimensional self-consistent methods on diatomic molecules. PhD Dissertation, The University, Helsinki, Finland.

Sundholm, D., Pyykkö, P., and Laaksonen, L. (1985) Two-dimensional, fully numerical molecular calculations. X. Hartree–Fock results for He_2, Li_2, Be_2, HF, OH^-, N_2, CO, BF, NO^+ and CN^-. *Mol. Phys.*, **56**, 1411–1418.

Thakkar, A.J. (1988) Higher dispersion coefficients: accurate values for hydrogen atoms and simple estimates for other systems. *J. Chem. Phys.*, **89**, 2092–2098.

Tillieu, J. (1957a) Contribution a l'étude théorique des susceptibilités magnétiques moléculaires. *Ann. Phys. Paris*, **2**, 471–497 (Parts I–III).

Tillieu, J. (1957b) Contribution a l'étude théorique des susceptibilités magnétiques moléculaires. *Ann. Phys. Paris*, **2**, 631–675 (Part IV).

Troup, G. (1968) *Understanding Quantum Mechanics*, Methuen, London.

Tunega, D. and Noga, J. (1998) Static electric properties of LiH: explicitly correlated coupled cluster calculations. *Theor. Chem. Acc.*, **100**, 78–84.

Van Duijneveldt-Van de Rijdt, J.G.C.M. and Van Duijneveldt, F.B. (1982) Gaussian basis sets which yield accurate Hartree–Fock electric moments and polarizabilities. *J. Mol. Struct. Theochem*, **89**, 185–201.

Van Vleck, J.H. (1932) *The Theory of Electric and Magnetic Susceptibilities*, Oxford University Press, Oxford.

Wahl, A.C. and Das, G. (1977) The MC-SCF method, in *Methods of Electronic Structure Theory* (ed. H.F. Schaefer, III), Plenum Press, New York, pp. 51–78.

Wheland, G.W. (1937) The valence-bond treatment of the oxygen molecule. *Trans. Faraday Soc.*, **33**, 1499–1502.

Wigner, E.P. (1959) *Group Theory and its Application to the Quantum Mechanics of Atomic Spectra*, Academic Press, New York.

Wolniewicz, L. (1993) Relativistic energies of the ground state of the hydrogen molecule. *J. Chem. Phys.*, **99**, 1851–1868.

Woon, D.E. and Dunning, T.H. Jr (1995) Gaussian basis sets for use in correlated molecular calculations. V. Core-valence basis sets for boron through neon. *J. Chem. Phys.*, **103**, 4572–4585.

Wormer, P.E.S. (1975) Intermolecular forces and the group theory of many-body systems. PhD Dissertation, Katholicke Universiteit, Nijmegen, The Netherlands.

Yan, Z.C., Babb, J.F., Dalgarno, A., and Drake, G.W.F. (1996) Variational calculations of dispersion coefficients for interactions among H, He, and Li atoms. *Phys. Rev. A*, **54**, 2824–2833.

Zener, C. (1930) Analytic atomic wave functions. *Phys. Rev.*, **36**, 51–56.

Author Index

Abbott, P.C., 245
Abramowitz, M., 40, 47
Aitken, A.C., 21, 23
Atkin, R.H., 37
Atkins, P.W., 197

Babb, J.F., 234
Babb, J.F., see Yan, Z.C., 225, 234
Bacon, G.E., 92
Ballinger, R.A., 111
Bartlett, R.J., 136
Battezzati, M., 244
Battezzati, M., see Magnasco, V., 244
Bishop, D.M., 234
Bonham, R.A., see Strand, T.G., 201
Born, M., 1, 142
Boys, S.F., 49
Briggs, M.P., 231
Brink, D.M., 44, 227
Buckingham, A.D., 237

Cade, P.E., 111, 112
Casimir, H.B.G., 223
Cencek, W., 153
Cheung, L.M., see Bishop, D.M., 234
Clementi, E., 111, 201
Cooper, D.L., 181

Corongiu, G., see Clementi, E., 111
Costa, C., see Figari, G., 234
Costa, C., see Magnasco, V., 153,
 154, 225, 231, 237, 239, 241, 244
Coulson, C.A., 164, 221, 240
Cybulski, S.M., see Bishop,
 D.M., 234

Dalgarno, A., see Yan, Z.C., 225, 234
Das, G., see Wahl, A.C., 135
Davies, D.W., 197
Dirac, P.A.M., 82, 86
Dixon, R.N., 77
Drake, G.W.F., see Yan, Z.C.,
 225, 234
Dunning, T.H., Jr., 110
Dunning, T.H., Jr., see Woon,
 D.E., 110

Eckart, C., 64
Ermler, W.C., see Rosenberg,
 B.J., 136
Eyring, H., 35, 42, 253

Figari, G., 224, 234
Figari, G., see Magnasco, V., 225,
 231, 237, 239, 241

Methods of Molecular Quantum Mechanices: An Introduction to Electronic Molecular Structure
Valerio Magnasco
© 2009 John Wiley & Sons, Ltd

Subject Index

Methods of Molecular Quantum Mechanics: An Introduction to Electronic Molecular Structure
Valerio Magnasco
© 2009 John Wiley & Sons, Ltd